Map-Seeking Circuits in Visual Cognition

Map-Seeking Circuits in Visual Cognition

A Computational Mechanism for Biological and Machine Vision

David W. Arathorn

Stanford University Press
Stanford, California

Stanford University Press
Stanford, California

Library of Congress Cataloging-in-Publication Data

Arathorn, D. W. (David W.)
   Map-seeking circuits in visual cognition : a computational mechanism for biological and machine vision / David W. Arathorn.
     p.   cm.
   Includes bibliographical references and index.
   ISBN 0-8047-4277-4
   1. Visual cortex. 2. Vision. 3. Computer vision. 4. Neural circuitry.
5. Cognitive maps (Psychology). I. Title.
QP383.15 .A736   2002
151.14–dc21   2002006116

Original Printing 2002
Last figure below indicates year of this printing:
11   10   09   08   07   06   05   04   03   02

This book was set in New Times Roman and Arial by the author and was printed and bound in the United States of America.

To North Fork

Contents

Figures

Figures

Foreword

Bruno A. Olshausen
*Associate Professor, Department of Psychology
and Center for Neuroscience, University of California, Davis*

Neuroscientists, psychologists, and engineers have long sought to understand how the brain transforms retinal images into meaningful representations of objects and surfaces in the environment. Among the most important unanswered questions is how one can recognize an object considering that its projection on the retina almost never appears exactly the same way twice. That this even presents a problem comes as a surprise to most, but when one thinks in neural terms about what needs to be explained it looms quite large. The late neuropsychologist Karl Lashley summarized the situation as follows:

> Visual fixation can be held accurately for only a moment, yet, in spite of changes in direction of gaze, an object remains the same object. An indefinite number of combinations of retinal cells and afferent paths are equivalent in perception and in the reactions they produce. This is the most elementary problem of cerebral function and *I have come to doubt that any progress will be made toward a genuine understanding of nervous integration until the problem of equivalent connections, or as it is more generally termed, of stimulus equivalence, is solved.* [1](emphasis added)

With this book, David Arathorn provides us with the first real glimpse of an answer to Lashley's dilemma. He has created a neural circuit that solves the correspondence problem—essentially, Lashley's "stimulus equivalence" problem—by dynamically modifying connections between input and output layers. He calls his network a map-seeking circuit, as it seeks to find a mapping between incoming data (images) and stored knowledge that provides the best match between them. The idea is actually quite general in that it could also be applied to correspondence problems involving comparisons across space or time, such as in stereopsis or motion. In fact, it seems quite possible that map-seeking circuits could be used in many parts of the cortex.

---

[1] Lashley, KS (1942) The problem of cerebral organization in vision, Biol Symp, 7: 301-322.

Dynamic remapping circuits were first proposed in 1947 by Walter Pitts and Warren McCulloch in order to explain how we recognize "universals," such as musical chords independent of pitch, or visual forms independent of size.[2] Since then there have been a number of other proposals along similar lines.[3, 4, 5] What sets Arathorn's scheme apart is his utilization of the "ordering property of pattern superpositions," which allows all remappings of the input and all stored patterns to be superimposed in the network during the map-seeking process. The best mapping between the input and a stored pattern is then quickly resolved via a recurrent computation in which connections between input and output layers are dynamically gated according to the degree of match between them. What is perhaps most impressive about this scheme is that it actually seems to work. Arathorn shows that the circuit is capable of recognizing objects embedded in fairly complex, natural scenes, and that the amount of time it takes to do so is consistent with physiological and psychophysical data on the timecourse of object recognition. A hardware implementation is currently in the works and is described in Chapter 8.

This work comes as a refreshing development, and it is perhaps no accident that it has emerged from outside the traditional academic circles in neuroscience and psychology. Arathorn is a computer scientist working primarily in industry who has had considerable experience devising practical solutions to complex problems. He has obviously thought long and hard about the problems of visual cognition and how to solve them using neural circuitry. What has emerged here is one of the most creative, insightful, and sophisticated computational models of visual cognition to date. For neuroscientists and psychologists, it should provide a framework for thinking about visual cognition in neural terms, something that is sorely needed in the design of experiments.

---

[2]Pitts W, McCulloch WS (1947) How we know universals: The perception of auditory and visual forms. Bulletin of Mathematical Biophysics, 9:127-147.

[3]Hinton GE (1981a) A parallel computation that assigns canonical object-based frames of reference. In: Proceedings of the Seventh International Joint Conference on Artificial Intelligence 2, Vancouver B.C., Canada.

[4]von der Malsburg C, Bienenstock E (1986) Statistical coding and short-term synaptic plasticity: A scheme for knowledge representation in the brain. In: Disordered Systems and Biological Organization (NATO ASI Series, Vol. F20) Bienenstock E, Fogelman Soulie F, Weisbuch G, eds., Berlin: Springer, pp 247-272.

[5]Olshausen BA, Anderson CH, Van Essen DC (1993). A neurobiological model of visual attention and invariant pattern recognition based on dynamic routing of information. The Journal of Neuroscience, 13, 4700-4719.

Only time will tell whether or not the brain actually uses map-seeking circuits. What we do know is that nature hides her secrets well, and it is only
through the construction and subsequent testing of models such as this one
that we will eventually unravel the inner workings of the brain.

Preface

Surprising as it may seem, the research reported in this book was not undertaken to discover a theory of visual cortical function. The author's original purpose was to use biological vision as guidance for the invention of a computational technology particularly suited to implementing visual functions. The investigation was predicated on the assumption that neuroscience and psychophysics had accumulated enough knowledge of the visual system to provide hints and constraints sufficiently specific to lead to a limited number of possible mechanisms. Using biology for guidance in invention is an old strategy, so one would not expect it to lead to a new place unless a new biological discovery or some other new principle is brought into the process. In this case the lynchpin proved to be a novel mathematical insight of the author's which had lain dormant for decades but now allowed the construction of a mechanism consistent with the evidence from the biological fields. This speculative "reverse engineering" led to a viable computational approach to a number of problems in vision which had defied practical solution, chief among which were the correspondence problem, and recognition under transformation. This had been the first goal of the effort.

Though the author's initial expectation was that the computational path would stray far from the biological evidence, the mathematical underpinnings in fact produced a mechanism that continued to conform remarkably to the biological constraints. As a result something in addition to a technology emerged as a by-product of the effort: a viable theory of cortical function. The author inadvertently found himself in the role of neurotheorist.

This book should be read, particularly by the neuroscience and psychophysics communities, with the inadvertence of the role in mind. Citations from the literature in those fields which appear in this book are only those which guided or confirmed the path of the investigation. Not having originated in a comprehensive review, they may ignore important and relevant contributions; the author apologizes to any authors who may have been unfairly overlooked.

The research described here was a solitary effort, with no institutional support or affiliation. Several individuals were influential. Important at the outset was Owen Chamberlain's encouragement of the author to undertake this voyage despite the certain high opportunity cost and uncertain outcome.

Crucial toward the conclusion was Grant Barnes' wider vision for the author's results, for without this there would have been no book and very likely no publication for other than a very narrow technological audience. Through Grant, who is Director Emeritus at Stanford Press, the research came to the attention of Ronald Bracewell, Charles Anderson and Bruno Olshausen, whose enthusiasm for the results buoyed the author's efforts. In addition to his contribution of a foreword, Bruno Olshausen's suggestions for improvements in the presentation of some of the material, particularly in Chapter 2, were gratefully incorporated. The author's gratitude is also due Norris Pope at Stanford University Press for his advocacy and support.

Chapter 1

Introduction to the Theory of Map-Seeking Circuits in Visual Cognition

1.1 Introduction

We should not be surprised that our capacity to see appears so dauntingly complex. The systems best understood by the human mind are those whose evolution was frozen longest ago: those of physics, chemistry, and molecular biology. Our habit has been to apply the principles that have simplified the comprehension of those ancient systems to more recently evolved systems, with the usual result that their apparent complexity remains undiminished. So it has been with vision. That, in the experience of the history of science, is a strong signal to start looking for a new principle.

This book introduces three interdependent concepts that prove to have great power in explaining the interpretation of three-dimensional visual inputs and recognition of patterns under transformation.

1. The starting assumption is that the vision process is decomposable into a flow of forward and inverse mapping operations. In many of these the application of data patterns at both ends results in the discovery of the correct mapping between them. This discovery mode of the mapping operation, termed *map-seeking*, solves the correspondence problem, which is the key to much visual processing.
2. An ordering property of pattern superpositions, apparently unrecognized until now, is the mathematical basis which allows map discovery, or *map-seeking*, to be resolved without combinatorial explosion.[1] Exploited in neuronal "hardware," this property permits the operation of map-seeking cortical circuits at speeds consistent with the performance and dynamics of biological visual systems.
3. The discovered mappings themselves, rather than the transformed pattern data, are often used as the inputs for subsequent stages of processing. The use of discovered mappings as data, termed *recharacterization*, is a basis for interpretation and generalization in the visual system.

---

[1] The author, Map-Seeking: Recognition Under Transformation Using A Superposition Ordering Property, *Electronics Letters*, 1 Feb. 2001, Vol 37 No 3.

Map-seeking is a fundamental, broadly applicable computational operation with algorithmic, neuronal and electronic implementations. Its generality makes it suitable as the core of an analytic method with which to approach cognitive functions. This book focuses primarily on its use as a basic module of the visual system. Determining the mappings between patterns, *map-discovery*, is essential at many stages of visual processing: from accommodating drift during fixations, to planning saccades, to stereopsis, to three-dimensional surface interpretation, to motion tracking, to assembling coherent representations from multiple fixations, to scene segmentation, to object and figure recognition, to repeated pattern location, to pattern abstraction and generalization, and very likely to processing steps not yet identified. But the application of map-discovery goes beyond vision. For example, it will be shown later to provide an elegant, unified solution to computing inverse kinematics for limb control in the presence of constraints and obstacles. And, though beyond the scope of this book, it has proven powerful in segmentation of phoneme sequences into words, and shows promise as a mechanism for classifying spatially mapped acoustic data into phonemes.

In short, the breadth of application of map-discovery suggests that it may be one of the basic functions of cortical circuitry. The central purpose of this book is to demonstrate some of that breadth in the area of vision and to demonstrate that plausible neuronal circuits consistent with neuroanatomy and neurophysiology are capable of performing the map-discovery function. A collateral purpose of the book is to propose a computing technology centered on map-discovery that promises to be far better suited to implementing cognitive functions than the current computing technology. Given the already central role of computing technologies in investigating brain function, an evolution of the technology toward the biology should assist the investigation of the latter.

The term *map-seeking* is used here to denote a class of mechanisms with a shared generic architecture whose function is the generalized result of *map-discovery*. Therefore this book makes two propositions: that map-discovery is an essential and widespread operation in vision and other brain tasks, and that map-seeking circuits are a general mechanism for executing that operation.

1.2 Map-Discovery in Vision

Anyone who thinks about the problem of vision and has watched a mountain goat bounding from ledge to ledge on a rock cliff can only marvel at the rapidity and accuracy with which the beast must construct a three-dimensional model of the terrain that will pass under its feet. The goat's manner of locomotion requires it, while a lizard scampering on the same cliff does not require a similar level of interpretive ability. He might have it, but he doesn't need it. Somewhere in the evolutionary line, perhaps when creatures began to move more rapidly on longer legs, when they had both the height to acquire an adequate view of the terrain ahead and the need to anticipate what their feet would encounter at the run, they developed the capacity to rapidly interpret two dimensional images of the terrain ahead into an internal three dimensional representation.

The study of vision, both natural and machine, has been largely directed at recognizing objects. But the visual mechanism which allows a mountain goat to leap as he does is not one that recognizes previously encountered objects but rather interprets three dimensional surface shape. Among the primary components of that interpretation are the parameters of the geometric transformation between displaced views of the same patch of terrain. The view displacement may be the distance between two eyes or a distance moved by the observer's head. In both cases the nature of the geometric transformation itself is data in the determination of surface shape, orientation, and often distance. This determination can be accomplished without specific recognition of any object in either of those images. A mate may be distinguished from a predator by color, vague size, smell, sound and general motion: attributes which do not require fine determination of three-dimensional form. Hoofed animals are not too particular about what spooks them. They may or may not be using complex visual transformations in the recognition of friend or foe. But the surefootedness of their flight makes it certain that they are recognizing the transformations in the images of their escape route and using these transformations to build the three dimensional terrain representation that let them land lightly and accurately at high speeds.

So, for the sake of the argument, let it be supposed that the transformation recognizing mechanism evolved into existence first for purposes of locomotion rather than recognition. This is a parsimonious supposition for a number of reasons. First, in the natural environment, terrain – and possibly the large vegetation upon it – may be the only immovable, rigid objects a creature encounters. Fewer and simpler transformations need be accommodated for terrain interpretation than for living object interpretation and recognition. Second, much less visual memory is required. Third, self motion increases the information available for terrain interpretation, while motion of another object adds complexity to the interpretation of the three dimensional shapes, particularly of deformable objects such as other creatures. Fourth, segmentation of scene into classified objects is not required for basic terrain interpretation. [2]

With this approach one defers taking a position in the long, unresolved debate over how three-dimensional objects are recognized. If recognition does sometimes resort to a three-dimensional model then interpretation must precede recognition for this mode at least. If this essential mechanism for terrain interpretation happened to evolve to be independent of the particular transformations accommodated, then its reuse in recognizing objects under transformation would simply be a matter of adding transformations. It would therefore embody a great step toward recognizing objects and interpreting their three-dimensional form.

The theory proposed here began with the conclusion of this argument: recognition of transformations precedes recognition of objects as a goal of the visual processing mechanism. This precedence may even reflect evolutionary precedence. The process responsible for this recognition must necessarily be concurrent, resolve quickly and be implementable with neural circuitry. These constraints led the author to the map-seeking mechanism as a general computational basis for a wide variety of visual capabilities. These capabilities include the obvious tasks such as three-dimensional interpretation, direct object recognition, scene segmentation, and motion tracking, but also a category of non-obvious tasks which may be accomplished deftly by recharacterization. This category includes recognition of figures or objects rendered with elements which are different from those with which the figure was rendered when originally learned. Recharacterization is an emergent behavior of map-seeking mechanisms, and gracefully explains a number of well-known human visual psychophysical behaviors.

The ordering property of image superpositions which underlies map-seeking permits the concurrent resolution of one or more stages of transformational mapping between an input image and one of possibly many images in memory. The generality and the breadth of application of this process make what is proposed here truly a theory and not a model. It is a theory of both biological and machine vision, proposed as an explanation of the former and a basis of development for the latter.

---

[2] All of this presumes that the correspondence problem can be solved without decomposing an image into objects, and this will be shown to be the case.

1.3 Context

Where does this theory fit in the body of facts and theories assembled by the various fields concerned with the mechanics of visual systems? Crudely one can partition those fields into three investigative domains.

1. behavior of biological vision
2. mechanisms of biological vision
3. mechanisms of machine vision

It will be noted that these are not the professional divisions, which often cross the boundaries. The first two domains are "scientific" in the traditional sense that they study a naturally occurring system. Experimental visual psychophysics provides the bulk of systematized knowledge about the behavior of biological vision. This knowledge forms a rich "functional specification" of the visual system which must be adhered to by any theory of the underlying mechanism. Investigations involving various modes of cell activity recording have yielded the most specific information about the mechanisms of biological vision, and though at present this information is not detailed enough to specify a mechanism, it does provide evidence that needs to be reconciled with any proposed mechanism.

Most of the work done in these two arenas is focussed on very specific phenomena and structures. Therefore the models which have been proposed to explain these phenomena have the same scope. The work reported here, in contrast, takes a "system view," in the sense that it provides a mechanism whose variants are capable of realizing a substantial span of the "functional specification" provided by visual psychophysics while remaining consistent with the constraints imposed by neurophysiology.

It is this breadth that makes the mechanism proposed here relevant to machine vision. Machine vision, with its pragmatic objectives, must perforce take a "system view." In general the field has made little effort to be guided by biological vision. There have been a number of reasons for this, most of them traceable directly or indirectly to limitations of hardware. While hardware capability continues to expand geometrically the legacy of earlier concepts linger. General-purpose machine vision remains elusive, and this cannot help but spark a longing to reverse-engineer biology's system, which for the foreseeable future will set the standard of performance.

Constructing a theory of biological vision and reverse-engineering the brain's visual system are congruent goals. Both uses are intended, and both will be discussed in this book.

1.4 Roots of the Theory

One of the obstacles to a "system level" theory of vision has been that the established repertoire of manipulations of visual data leads to impossibly large allocations of neural resource to achieve the observed speed of interpretation and recognition. A primary contributor to this problem of "combinatorial explosion" is the variety of transformations to which the view of a target is subject. The persistent inability of any of the vision science fields to make a dent in this problem suggested to the author that a unifying mechanism for vision must be based on at least one new property. That new property, went the reasoning, would make the correspondence problem[3] simple, since this lay at the heart of all the higher level processing. A robust solution to the correspondence problem precluded an approach commonly encountered in both vision theory and machine vision: first establishing "anchoring" matches between specific features of a input image and a memorized image and then using these to determine the transformations between the two images. Anchoring is impractical for at least two reasons. In the case of the terrain problem, which introduced this discussion, the scene often fails to present sufficiently unique anchoring features. And, if anchoring features could be found they would have to be unique in geometric-transformation-invariant attributes such as color, since shape characteristics are subject to the very transformations the system is trying to determine.

A classic set of experiments in visual psychophysics provides compelling evidence that the correspondence problem is solved at least in one part of the visual system without recourse to anchors: the random dot stereograms of Bela Julesz.[4] For even if one assumes (implausibly) that the initial vergence is hunted down in a dot by dot scan, some piece of neuronal machinery almost instantaneously computes the translation (and usually perspective disparity) between the two views of the second virtual plane. Given nature's propensity to reuse its inventions, one might expect that machinery to be closely related to the machinery for interpreting shape from motion. The latter must match images from two moments instead of matching images from two eyes, but it must accommodate a far greater range of possible transformations than the machinery for stereo disparity.

What sort of neuronal process could result in such a rapid determination? A likely candidate might be some form of competition with feedback to accelerate the selection process. What must be selected, continued the logic, was the most computationally useful entity: the correct transformation. Therefore a number of transformational mappings between the input image and a

---

[3] Marr, p188ff.
[4] Julesz 1960

memorized image must compete based on the quality of the match between one of the images and each mapping of the other.

So far this concept would not seem to lie outside any traditional boundary of method or property and a number of theorists [5, 6] have in fact moved in this direction. But all solutions based on traditional principles are blocked by combinatorial explosion problems or have moved down a path which does not yield the piece of information necessary for terrain interpretation: the explicit measure of the transformation between the two images.[7]

Combinatorial problems arise in two areas. First, the number of possible visual transformations required by the simplest terrain problem is substantial: a range of scalings times a number of rotations in the plane times a number of rotations into the plane times the number of possible translations necessary to register the two images. When the problem is extended to recognition of rigid objects a wider range of scalings, rotations and perspective transformations is required. The problem can only be solved by a mechanism that allows composition of transformations: for example, a stage of translation and scaling followed by a stage of rotations on and into the plane. To keep such a solution from exploding combinatorially in the time dimension these two stages have to resolve their competitions concurrently and cooperatively to the correct composite transformation.

Second, with only a single copy of the input image and a single copy of the memory image, separate pathways between them for each mapping would seem to be required for the competition to proceed concurrently. Or each possible composed mapping would have to be tested sequentially. If the problem is extended to object recognition there must be multiple memory candidates, and the resources in time or material necessary to establish the mapping between input and one memory would have to be multiplied by the number of candidate memories.

The obvious way to conserve resources is to let all the mappings communicate concurrently with memory via the same conduit or channel. Multiplexing that channel is precluded since it would simply push the combinatorial explosion into the time domain. The only conclusion is that the channel carries a superposition of the signals created by each mapping and somehow the correct component of the superposition is isolated.

---

[5] Marr, Poggio 1977, 1979
[6] Van Essen, Anderson, Olshausen 1994
[7] Lades, Vorbruggen, Buhmann, von der Malsburg, Wurtz, Konen 1993

At this point inference pointed to the existence of some ordering property of image superpositions which could be exploited by a competition based on the quality of match between one image and each transformational mapping of the other. A bit of math and a little computational experiment confirmed the existence of a property with the hoped-for characteristics. Immediately it was obvious it could be exploited for superpositions of memory responses as well as superpositions of mappings of the input. The competitive mechanism could be used to prune the contributors to one of the superpositions and the whole process would converge to the correct mapping and correct memory response. A reciprocal pathway architecture was clearly indicated, but of course what else would one expect from the evidence of massive reciprocity in the pathways of the visual cortices?

The first experimental circuit, a single-layer neuronal architecture constructed along these lines, demonstrated that the correspondence problem was robustly solved by this competitive concurrent search for mapping and memory. *Map-seeking* seemed a fitting description of the process. But with a single layer combinatorial problems still nagged which, as mentioned earlier, could only be solved by deploying composable mappings in multiple layers. But if the circuit was capable of resolving superpositions from memory, why shouldn't it be able to resolve superpositions coming back from another layer of circuitry? This proved to be the case. And it turned out that solutions to other seemingly disparate vision behaviors were intrinsic to the circuit. Map-seeking appeared to be a unifying principle for vision. Now the task was to see how wide a span of vision problems it could solve and what its limitations were, and to determine if it was compatible with neuronal implementation. Those are the primary subjects of this book.

It must be borne in mind that the map-seeking principle operates on a property that is independent of implementation. Two implementations, one algorithmic and one "neuronal," will be presented and their equivalence will be demonstrated. These are certainly not the only algorithmic and neuron-like implementations possible. The principle is not tied in any way to the details of particular models. The instances used for illustration here are chosen for simplicity and have in no way been optimized for speed or capability. The general philosophy comes from engineering: a design must work pretty well without optimization if it's expected to work very well with optimization.

Despite the simplicity and elegance of the algorithmic implementation it is not being proposed here that the visual or other cortices have evolved to execute this algorithm. An algorithm is a sequence of mathematical or logical operations designed to achieve a certain computation, and it would be a curious inversion to suppose that biological machinery has evolved to execute, even approximately, mathematical operations. Rather, the mathematics of the algorithmic circuit roughly approximate the emergent behavior of certain arrangements of non-linearly interacting elements: synapses, dendritic branches, neurons, circuits of neurons. It is only a reflection of our mathematical techniques and current digital computing machinery that this approximate form should be easier to analyze and quicker to simulate. Had analog computation survived as a viable technology, the simulations of the neuronal forms of the circuit could now be executed far more quickly than the digital implementations of the algorithmic circuit, using far fewer transistors running at much lower frequencies, as will be seen in later chapters.

1.5 Scope and Organization

One purpose of this book is to present the map-seeking principle and a range of biological visual and motor capabilities for which it provides a unified mechanism. Another purpose is to introduce the map-seeking principle as a basis for engineering solutions in machine vision and possibly other areas of perceptual and cognitive computation. In biological and engineered systems a basic principle may be expressed in many ways, and with endless refinements and additions to extend its powers. This book presents only two expressions of the principle. Explorations of refinements or additions lie outside its scope. Consistent with this approach, all the demonstrations contained in this book are performed by algorithmic or neuronal circuits which are pure, minimal expressions of the principle. They are not demonstrations of what is possible in applications of the principle refined by evolution or engineering.

The work presented here will address a wide range of interests; its findings bear directly on neuroscience, cognitive psychology and machine vision. Unfortunately these fields do not yet share common mathematical approaches or technical constructs. Therefore the simplest possible mathematical representation is used to present the ideas. Most technically detailed discussions are isolated either in Appendices or in sections which may be skipped or revisited at any point.

Chapter 2 presents an explanation of the structure and operation of map-seeking circuits. An algorithmic form of the circuit, constructed entirely of simple operations on vector variables, is most amenable to discussion and analysis. The dataflow pathways and the "pure" mathematical operations of this algorithmic implementation have very close counterparts in neuronal and analog semiconductor map-seeking circuits. The behavioral equivalence between algorithmic and neuronal implementations is demonstrated.

Chapter 3 investigates the application of the circuits to terrain interpretation and construction of viewpoint-independent 3D models of terrain or objects. The problem is shown to utilize both the *map-seeking mode* and the *driven-mapping mode* of the circuit. In the former mode the circuit discovers the mapping between two sets of data, and that mapping itself becomes the data for recharacterization and other further computations. In the latter mode one mapping is selected by an external signal and used to transform an input for capture or some further computational purpose.

Chapter 4 ties visual space to kinematic space, completing the path between eye and limb using map-seeking circuits at every stage. The application of the *map-seeking mode* and the *driven-mapping mode* of the circuit to inverse and forward kinematics is discussed and demonstrated. A sketch of computational architecture for 3D interpretation and limb positioning using a number of map-seeking circuits is presented.

Chapter 5 discusses the architecture and dynamics of the neuronal implementation. The connectivity and dynamics of the neuronal circuit are related to the algorithmic implementation presented in Chapter 2. Aspects of the neuronal circuit which closely reflect biological cortices are discussed, including reciprocal nature of the pathways, oscillatory dynamics, and arborized non-linear dendritic structures. Aspects in which the circuits diverge from biological realism are also discussed.

Chapter 6 focuses on recognition, segmentation, motion tracking and recharacterization. Most of the demonstrations in Chapter 4 use the oscillatory neuronal implementation. The biological plausibility of the circuit extends to real-time performance, and this is demonstrated in terms of oscillatory periods in order to relate it to cortical performance.

Chapter 7 discusses specific human visual neurophysiological and psychophysical behaviors readily explainable by the function of map-seeking circuits. Most of these have remained without any computationally explicit explanation until now. The parsimony of explanations of these behaviors based on map-seeking is strong support for its candidacy as a theory of human visual function. Other support comes from malfunctions of the circuit that can be induced by mistuning its dynamics. These mimic visual deficits associated with schizophrenia and may point to their cause.

Chapter 8 discusses technological uses of map-seeking circuits, both as embedded components and as the basis of large scale computing platforms for cognitive applications.

Chapter 9 discusses lines of further research suggested by the work presented here.

Chapter 2

The Algorithmic Map-Seeking Circuit

2.1 Relationship to the Theory

The purely numeric mechanism to be presented in this chapter reveals aspects of the operation of map-seeking circuits that are much more difficult to analyze in the more biologically realistic neuronal circuit which will be discussed starting in Chapter 5. Nothing resembling a neuron will be found in this chapter, yet the behavior of this numerical mechanism and the behavior of a circuit composed of cell-like entities with detailed dendritic structures are remarkably similar, as will be seen. The similarity is due to a common structural organization and a common sequence of operations on the signals, despite the difference in the medium that carries those signals.

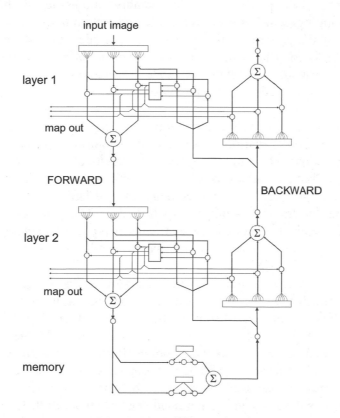

Fig. 2-1: Multilayer circuit.

The common elements of all map-seeking architectures are one or more layers of paired forward and backward signal pathways. Each pathway is a bundle of individual signal lines, enough to carry a spatially organized pattern such as an image. Within each layer the spatial organization of the individual signal lines in the bundle may be rearranged in a variety of ways between the input and the output of the layer. The convention will be to consider each pathway bundle to be an ordered set of signals represented by a vector of signal amplitudes. Each possible rearrangement between the input and the output bundle is a mapping between those sets or vectors. Within each layer there is an inverse mapping on the backward path for each mapping on the forward path. The mappings are therefore part of the "wiring" of each layer. For circuits used in vision the mappings correspond to the visual transformations translation, rotation, scaling, and perspectivity.

A circuit with two mapping layers and a memory layer can be seen in Fig. 2-1. (The term "layer" here is not meant to imply a layer in cortical anatomy.) The functional goal of the circuit is to discover the best sequence of forward mappings which transform patterns in the data in the input field to match one of the patterns stored in memory, and associated inverse backward mappings which transform the selected pattern in memory to match all or part of the pattern in the input field. The identities of the discovered mappings in each layer are as important an output as the transformed images. The circuit can operate in a mode in which one or more mappings are selected or weighted by external inputs.

In vision the geometric transformations produced by changes in spatial relationship of viewer and object are decomposable into sequences of simpler transformations. The multiple layers of the circuit allow a large repertoire of aggregate transformations to be composed from much smaller repertoires in each layer. In circuits that will be demonstrated later, layer 1 contains several thousand translation mappings while layer 2 (and in one case layer 3) contains hundreds or thousands of rotated perspective mappings. The aggregate repertoire is hundreds of thousands or millions of transformations.

The discovery of aggregate mappings is a problem faced not just by the visual system, but by the motor system in solving the inverse kinematics of limbs. The circuits described here are also efficient at solving inverse kinematics, as will be demonstrated in a later chapter. In these three-layer circuits, each layer contains the mappings from a joint to the end of a limb segment. When presented with the target position in the input field the map-seeking process discovers the sequence of mappings that position all three limb segments so the end of the hand segment touches that point, obeying joint range constraints and avoiding obstacles if present. It is very likely that the map-seeking mechanism is also applicable to speech recognition and other cognitive processes.

Discovering the best sequence of composable mappings would be trivial if the speed of the mechanism allowed each combination of mappings and memory patterns to be tested sequentially. But neither biology nor technology offers any medium fast enough to overcome the combinatorial explosion that occurs when practical numbers of mappings and memories are involved. Nor can sufficient speed be gained by creating separate paths for testing combinations in parallel, since the combinatorial explosion is simply moved from time to transistors or neurons.

Instead, the map-seeking circuits described in this book employ a more subtle process of selecting the best mappings. It exploits a property of pattern superpositions (signal summations) which allows an aggregate mapping and memory pattern to be selected in a process which converges largely independent of the number of mappings and memory patterns being tested. As a consequence, the investment of material resources grows additively with the number of individual mappings deployed in all layers and number of patterns in memory, rather than multiplicatively with the number of aggregate mappings and memory patterns which may be composed.

The superposition ordering property is inherent in map-seeking circuits implemented with a variety of mediums. We will first examine a minimal numeric implementation, since it can be depicted and analyzed most simply. In later chapters we will examine both neuronal and electronic implementations which exploit the same property.

2.2 Elements of the Algorithmic Circuit

All circuits are composed from a few basic operations on sets or vectors of signals: *map*, *combine*, *match*, *attenuate*, *scale*, and *compete*. The overall function of the circuit is rather insensitive to the particular characteristics of the implementation of these operations. In this chapter the simplest arithmetic implementations will be used. Because of their terseness and familiarity a circuit built from these operations can be depicted compactly and readily understood. The operations will seem too pure to have much significance in neuronal circuitry, but it will be seen later that structurally identical circuits built with non-linear versions of each operation have remarkably similar behavior. The essential characteristic of each operation type is monotonicity rather than linearity or a particular form of non-linearity.

The flow of numerical operations presented here, though circular, if standing alone would properly be called an algorithm. However, the term *algorithmic circuit* is used throughout this book to emphasize that it is a particular implementation of a more universal structure: an isomorphism of the neuronal and electronic implementation which are truly circuits in the conventional sense.

In vision circuits the vector representing a signal should be visualized as a two-dimensional array (not necessarily Cartesian) representing the presence of image features such as edges. Such a signal might be produced by applying edge filtering to a continuous image. This 2D spatial interpretation of the vector means that visual transformations such as translation, rotation, scaling and perspectivity can be applied to vector $\mathbf{v}_1$ by selecting and rearranging its elements into the elements of vector $\mathbf{v}_2$. A recognizable pattern represented by vector $\mathbf{v}_1$ would appear in vector $\mathbf{v}_2$ transformed by some combination of translation, rotation, scaling and perspectivity. The terms *map* and *mapping* refer here to the rearrangement of one vector to form another to implement such visual transformations.

The operations fall into three groups:

arithmetic operations

$$\mathbf{v}_1 + \mathbf{v}_2 \rightarrow \mathbf{v}_3$$ *combine* vectors $\mathbf{v}_1$ and $\mathbf{v}_2$ additively element by element

$$\mathbf{v}_1 \bullet \mathbf{v}_2 \rightarrow x$$ *match* determines a scalar measure of similarity x between vectors $\mathbf{v}_1$ and $\mathbf{v}_2$

$$k \cdot \mathbf{v}_1 \rightarrow \mathbf{v}_2$$ *attenuate* all elements of vector $\mathbf{v}_1$ by factor k
or $\quad k \times \mathbf{v}_1 \rightarrow \mathbf{v}_2$

mapping operations

$$d(\mathbf{v}_1) \rightarrow \mathbf{v}_2$$ *map* all or a subfield (a set of elements corresponding to a contiguous 2-dimensional area in the input image) of vector $\mathbf{v}_1$ onto vector $\mathbf{v}_2$ to implement some spatial transformation

$$d^{-1}(\mathbf{v}_2) \rightarrow \mathbf{v}_1$$ *inverse* of *d( )* map vector $\mathbf{v}_2$ into vector $\mathbf{v}_1$ (used in multilayer circuits and attention-shifting)

scaling and non-linear operations

$$s(\mathbf{v}_1) \rightarrow \mathbf{v}_2$$ *scale* all elements of vector $\mathbf{v}_1$ proportionally such that the largest element of $\mathbf{v}_1$ is no greater than unity

$comp(\mathbf{v_1}, \mathbf{v_2}) \rightarrow \mathbf{v}_1$ *compete* among elements of vector $\mathbf{v_2}$ by attenuating all but the largest element of $\mathbf{v_2}$ proportional to the difference between each element and the largest element, then use these to modify the corresponding values of $\mathbf{v_1}$

$f(x)$ a scalar non-linear monotonic *gain* function either convex or sigmoidal depending upon application

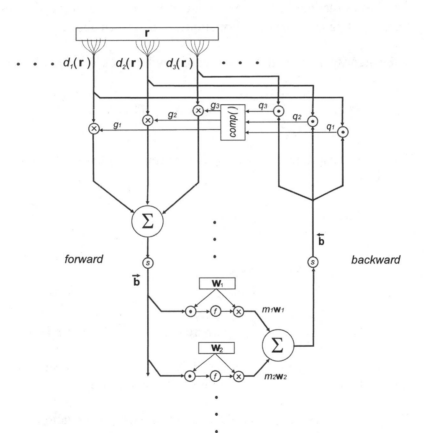

Fig. 2-2: Dataflow representation of map-seeking algorithm. Signal paths are vector (indicated with a heavy line and bold symbol) or scalar (indicated by light line and italic symbol). Each bundle of scalar paths emerging from the rectangle labeled **r** designates a mapping $d_i(\mathbf{r})$ of all or part of the vector **r** onto the vector pathway emerging from the bundle.

Fig. 2-2 shows a single layer circuit with memory. A single layer circuit is used for simplicity, but the principles apply equally to multilayer circuits, as will be discussed later. The operations listed above are executed in the sequence shown in Fig. 2-2.

In the algorithmic circuit used throughout this book, the operation symbols $+$, $\bullet$, and $\times$ are interpreted to have conventional meanings: vector addition, vector dot product, and scalar-vector multiply (not cross product).

The mappings and inverse mappings $d(\ )$ and $d^{-1}(\ )$ belong to a repertoire of re-orderings (one-to-one, without interpolation in the simplest case) corresponding to visual transformations, as described above.

In the algorithmic circuit demonstrations in this book $f(\ )$ is implemented as

$$f(x) = \begin{cases} x^n & if \ x \geq t \\ 0 & if \ x < t \end{cases} \quad where \ n > 1, 0 < t \ll 1$$

The competition function $comp(\ )$ can be implemented in a variety of ways including[1]

$$comp(\mathbf{v}_1, \mathbf{v}_2) = \max\left(0, \ \mathbf{v}_1 - k \cdot (h - \mathbf{v}_2)^n\right)$$

$$h = \begin{cases} \max \mathbf{v}_2 & if \ \max \mathbf{v}_2 > t \\ t & if \ \max \mathbf{v}_2 \leq t \end{cases}$$

$$where \ n \geq 1, 0 < k < 1, 0 < t \ll 1$$

Notice that if the greatest element of $\mathbf{v}_2$ is less than a threshold t, $comp(\ )$ pushes all the values of the resulting vector toward zero.

The $comp(\ )$ function is used in the circuit as recurrence relation.

$$\mathbf{v}_1' = comp(\mathbf{v}_1, \ \mathbf{v}_2) \quad where \ \mathbf{v}_1' \ is \ value \ of \ \mathbf{v}_1 \ at \ time \ t+1$$

---

[1] For all scalar-vector arithmetic operations the scalar is applied to all elements of the vector, so the result is a vector.

The scaling function *s* adjusts the signal vectors so that their magnitudes remain in approximately the same range regardless of the number of superimposed components contributing to each.

$$s(\mathbf{v}) = \begin{cases} \mathbf{v}/\max \mathbf{v} & \textit{if } \max \mathbf{v} > 1 \\ \mathbf{v} & \textit{if } \max \mathbf{v} \leq 1 \end{cases}$$

The purpose of this scaling is to make consistent use of the non-linearities in *comp( )* and *f( )*.

In the equations below, to eliminate multiple subscripts, the signal on a path is denoted by the letter under an arrow indicating path and direction: right/left pointing arrow denotes forward/backward path. e.g. $\vec{\mathbf{b}}$ denotes the signal on path b_{fwd}, $\overleftarrow{\mathbf{b}}$ denotes the signal on path b_{bkwd}. The scaling functions *s( )* are omitted in the equations below to simplify the expressions. (Recall that *s( )* does not alter the relative magnitudes of vector elements and therefore does not alter the ordering of the results of matching operations.) Vector quantities are in bold, scalar in italic.

The input field path to the circuit is designated r_{fwd}, (for *receiver*), carries the signal $\vec{\mathbf{r}}$. In vision applications the input field typically consists of the output of thousands of feature filters or detectors. Subfields of $\vec{\mathbf{r}}$ are transformed (spatially) by a set of mappings $d_i( )$ to produce a set of vectors each of which is attenuated by an associated coefficient g_i. For vision typical sets of mappings are those implementing translation, rotation, scaling, perspectivity, rotations into plane, and so forth.

The superposition on path b_{fwd}, (for *bus*) is $\vec{\mathbf{b}}$ which is the sum of mapped subfields, $d_i(\vec{\mathbf{r}})$ each attenuated by g_i. So the aggregate signal $\vec{\mathbf{b}}$ is

$$\vec{\mathbf{b}} = \sum g_i \cdot d_i(\vec{\mathbf{r}}) \qquad\qquad \text{eq. 2-1}$$

(The coefficients g_i are initially set to unity, but will be adjusted later in the signal flow.)

The signal on b_{fwd}, $\vec{\mathbf{b}}$ is the input to memory. Each memory unit has paired forward and backward paths m_{fwd} carrying the scalar $\vec{m}$, and m_{bkwd} carrying the vector $\overleftarrow{\mathbf{m}}$. Each memory unit contains a vector of weights, **w**, which represents the pattern learned by that memory unit. (The weights for each memory unit have been set by applying the target pattern to b_{fwd} and setting the weights so that $\vec{\mathbf{b}} \bullet \mathbf{w}_k = 1$. See end of Appendix E.)

The degree of match between each weight vector and the signal on b_{fwd} is computed with a dot product, producing a scalar value to which the non-linearity $f(\ )$ is then applied.

The signal on each m_{fwd} path is the scalar

$$\vec{m}_k = f\left(\vec{\mathbf{b}} \bullet \mathbf{w}_k\right)$$

eq. 2-2

The resulting scalar is used as a coefficient to attenuate the values of the weight vector $\mathbf{w}_k$ to produce a vector which is simply the weight vector or memorized pattern scaled by the degree of match between the b_{fwd} signal and that vector of weights. The signal on each m_{bkwd} path is the vector

$$\bar{\mathbf{m}}_k = \vec{m}_k \cdot \mathbf{w}_k = f\left(\vec{\mathbf{b}} \bullet \mathbf{w}_k\right) \cdot \mathbf{w}_k$$

eq. 2-3

The superposition on path b_{bkwd} is the sum of the attenuated weight vectors, $\bar{\mathbf{m}}_k$. So the aggregate signal is

$$\bar{\mathbf{b}} = \sum_k f\left(\vec{\mathbf{b}} \bullet \mathbf{w}_k\right) \cdot \mathbf{w}_k$$

eq. 2-4

The pairing of forward and backward paths is characteristic of all implementation of map-seeking circuits. Just as b_{fwd} carries a superposition of signals produced by the mappings of the input field, so b_{bkwd} carries a superposition of signals produced by memory. In both of these superpositions the components are scaled by coefficients produced by measuring the degree of a match between two patterns. In the b_{bkwd} superposition the coefficient of each component vector is $\vec{m}_k$. In the b_{fwd} superposition the coefficient of each component vector is g_i. Shortly we will see how g_i is determined.

The signal on b_{bkwd} follows two sets of paths. In one set of paths the signal is compared to each mapped subfield $d_i(\vec{\mathbf{r}})$ of the input field. These are the vectors that were earlier multiplied by the corresponding g_i and summed to form the signal on b_{fwd}. The comparison is computed as a dot product of the two vectors to produce an intermediate goodness-of-match signal, the scalar

$$q_i = \bar{\mathbf{b}} \bullet d_i(\vec{\mathbf{r}})$$

eq. 2-5

The vector comprising all of the q_i is designated **q**. A competition function, *comp( )*, is applied to the vector of all intermediate goodness-of-match signals, **q** to increase the difference between the greatest of these and all the others and produce a vector of final scalar goodness-of-match signals g_i. The vector comprising all of the g_i is designated **g**. The new value of g_i is computed as a function of its last value and the new value of q_i.

Therefore *comp( )* is implemented in practice as a function both **q** and **g**. The new values of g_i and **g** are denoted g_i' and **g**′.

$$\mathbf{g}' = comp(\mathbf{g}, \mathbf{q})$$
eq. 2-6

As discussed earlier *comp( )* is a monotonic, non-linear function which preserves the magnitude of the largest competitor and diminishes all others so that repeated application ultimately leaves the largest competitor at its original value and leaves all others at zero. For some purposes it is useful to construct *comp( )* so that a number of winners will survive if they begin within a certain small range of the largest. For this purpose *comp( )* is built around a sigmoid-like function.

The other pathway for the b_{bkwd} signal is the paired pathway of the input field. This pathway is not needed to understand the operation of a single layer circuit, so it will be discussed later in the context of multilayer circuits.

Once **g**′ is computed the iteration is complete. The new values of the vector **g**′ are now applied as the coefficients of the mapped subfields of the input $d_i(\mathbf{\bar{r}})$ to alter each one's amplitude as a component of the new superposition signal on b_{fwd}. The cycle continues as described above with the new composition of the b_{fwd} signal, which in turn alters the responses of each memory unit, thereby adjusting the amplitudes of the components of the superposition signal on b_{bkwd}. This in turn alters the values of the intermediate goodness-of-match signals **q**. The differences in **q** are amplified by *comp( )* and yield the new, more highly differentiated values of **g**.

The process converges when the vector **g** contains only one non-zero element (or a few in the case of repeated patterns). As a consequence only one mapped subfield $d_i(\mathbf{\bar{r}})$ of the input is contributing to the signal on b_{fwd}. For only one element of **g** to be non-zero the signal on b_{bkwd} must substantially match the mapped subfield $d_i(\mathbf{\bar{r}})$. For this to be the case, the weight vector

$\mathbf{w}_k$ of at least one memory unit must significantly match $d_i(\vec{\mathbf{r}})$ and the associated coefficient $\vec{m}_k$ must be significantly more than zero. But if $\mathbf{w}_k$ does significantly match $d_i(\vec{\mathbf{r}})$, which is the only component on b_{fwd}, then $\vec{m}_k$ will naturally be close to unity. This is the stable end state of the convergence when some part of the input field can be mapped to form a substantial match with some pattern store in a memory unit. If no such match can be achieved the entire $\mathbf{g}$ vector goes to zero and no signal is present on b_{fwd}, or elsewhere in the loop.

The ordering property of superpositions mentioned in the first chapter is responsible for the convergence settling on the correct element of the mapping coefficient vector $\mathbf{g}$. Stated simply, ***when matches are computed between a superposition and a set of patterns, the degree-of-match will, with high probability, be the greatest for that pattern which has the closest match to one of the components of the superposition***. The probability of this being true is inversely proportional to the number or components contributing to the superposition. Therefore it can be unreliable to chose the winner on the basis of the first round of matching. Instead, the number of contributors to the superposition is reduced gradually, thereby steadily increasing the probability that the greatest degree-of-match belongs to the pattern with the closest match among the remaining superposition components. (In order for this process to work in practice the superposition signal must not saturate the channel that carries it. This imposes a restriction on how the signal can be implemented in analog systems, as will be discussed in Chapter 5.)

In the process of convergence all mappings $d_i(\vec{\mathbf{r}})$ start out contributing to the b_{fwd} superposition. Also all memory patterns $\mathbf{w}_k$ start out contributing to the b_{bkwd} superposition. But as iterations progress increasing numbers of the contributors to both drop out because of poor matches to any of the remaining components (or combinations of components) of the superpositions with which they are being matched. Therefore as the convergence proceeds the probability increases that the element of $\mathbf{g}$ with the highest value will correspond to the best match between input field and a memory pattern.

It is possible for unusual combinations of contributors to the superpositions, whimsically termed *collusions*, to cause the incorrect element of $\mathbf{g}$ to assume a higher value than the correct element of $\mathbf{g}$. Normally this ordering reverses as the number of contributors to the superpositions diminishes, as seen in Fig. 2-12 iterations 1-25. If, however, it persists long enough for *comp( )* to drive the correct element of $\mathbf{g}$ below threshold, the circuit will proceed to converge to the non-recognition state: signaling it has found no matches between the input field and memories. This happens because the incorrect elements of **g** by definition do not correspond to a mapped subfield $d_i(\vec{r})$ which has any substantial match with a single memory, and therefore will also be suppressed by either the memory threshold or the mapping threshold. (If one of those elements of **g** did correspond to a mapped subfield with a match in memory it would constitute a correct convergence.) The history of $\mathbf{m}_{fwd}$, **q** and **g** prior to the final state reveals the weak matches contributing to the collusion. Because the evidence of these weak matches is not normalized away there are a number of approaches to recovery from a collusive convergence.[2] A more complete discussion of the ordering property and conditions for correct convergence appears later in this chapter.

The non-linearity $f(\,)$ in eq.( ) assures that only a significant match between $\vec{\mathbf{b}}$ and $\mathbf{w}_k$ will result in a substantial coefficient $\bar{m}_k$ and consequently a significant presence of the pattern $\mathbf{w}_k$ in the b_{bkwd} superposition. In practice $f(\,)$ contains a threshold. If the patterns stored in the various memory units are largely different, then only one memory unit will finally be responding when the circuit converges during a successful recognition. If the patterns stored in memory are quite similar, several might yield non-zero responses to the same signal on b_{fwd}. If this behavior is undesirable, all but one of those responses can be eliminated by applying the *comp( )* function to the vector of $\bar{m}_k$ memory responses and using the resulting, more highly differentiated vector as the coefficients of the components of the b_{bkwd} superposition. A circuit of this configuration is demonstrated later in the chapter.

If there are several patterns in the input field that, mapped, correspond to patterns stored in memory, usually only one of these will be matched when the process converges. This is because the *comp( )* function assures that only one non-zero element of **g** will survive. If several patterns in the input field closely match a single memory several non-zero elements of **g** will survive if *comp( )* is not too aggressive. The non-zero elements of **g** will then mark the presence of the occurrences of these repeated patterns. This is demonstrated in Chapter 6.

A variant of the circuit, which will be demonstrated later in this chapter and elsewhere, uses the b_{bkwd} signal to suppress the matching pattern in the input field after the first convergence. This allows any other patterns in the input

---

[2] These generally involve altering the initial weighting of mappings or memories involved in the collusion and starting again with the same input.

field to find a match in memory on subsequent convergences. This series of convergences in effect segments the input field in a series of "attention shifts."

2.3 Behavior of Two Circuit Implementations

In the preceding discussion it was emphasized that the process of convergence is independent of the particular implementation of the circuit, and that the basic operations listed earlier can have very different characteristics without disturbing the process. This can be effectively demonstrated by using the same data with two very different implementations: an algorithmic circuit based on the equations presented above, and an analog "neuronal" circuit whose computational elements are non-linear and more similar to neurons or transistors, and whose signals are periodic pulses representing relative signal strengths by phase differences rather than numbers. The analog or "neuronal" circuit is discussed at length in Chapter 5.

The equivalence is demonstrated in a task involving matching a target pattern somewhere in the input field, under transformation, to one of multiple memory candidates. Though the task involves distinguishing identity (i.e. which memory) it is not intended as a realistic recognition task, at least in the biological sense. [3] This simple capability has many uses in machine vision, but in the context of biological vision it should be considered only a primitive operation from which a biologically realistic process of recognition is built. A demonstration of a more realistic process involving "recognition by parts" using a three-layer circuit is found in Appendix I.

The task here is to identify one of a number (up to thousands) of memorized patterns which matches a target section of a test input image. Test patterns and memory patterns are produced from the photograph of the tiger, Fig. 2-3(a), by Rothwell edge filtering.[4] A 90 x 90 section[5] of the full edge-filtered image, Fig. 2-3(b), taken at random from the area of the tiger (to simulate a saccade) provides the test input images, Fig. 2-3(c,d). A 40 x 40

---

[3] "Recognition" as practiced by an animal in a natural environment is a complex, and at present an under-defined, process. Evidence for a piecewise process of recognition is discussed in Chapter 7 and demonstrated in Appendix I.

[4] Photo and Rothwell edge data courtesy Heath, Sarkar, Sanocki, Bowyer (1997).

[5] The image size restriction was imposed by physical computer memory limitations affecting the neuronal simulation. The same constraint precluded large numbers of memories and large numbers of mappings in the neuronal simulation, and consequently in the algorithmic circuit to demonstrate equivalence.

image taken at random along the edge of the unrotated test image serves as the memory containing the target pattern, Fig. 2-3(e).

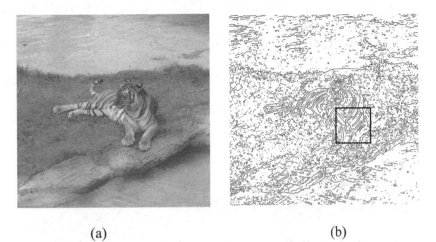

(a)   (b)

The data input fields below are presented inverted relative to (a) and (b)

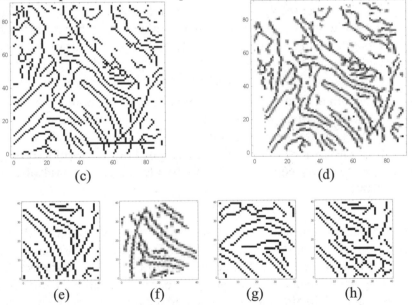

(c)   (d)

(e)   (f)   (g)   (h)

Fig. 2-3: Training and input data. (a) source image; (b) Rothwell edge detection applied to (a), square shows test image area; (c) 90x90 test image 1 – dark line shows location of lower edge of target area; (d) test image 2: 10 degree rotation of (c); (e) 40x40 memory pattern of target area; (f) 160 degree rotation of (e); (g) example distractor memory; (h) example of permuted tile distractor memory.

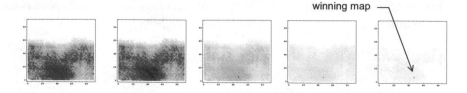

(a) r-match cell values (equivalent to **g**): convergence of neuronal circuit for test image 1.

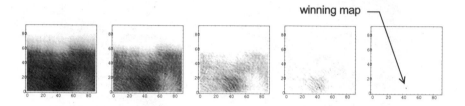

(b) **g** values: convergence of algorithmic circuit for test image 1.

Fig. 2-4: Test results. Two sequences of steps in the convergence of (a) a neuronal circuit and (b) an algorithmic circuit, both using same input data. Each frame is a density plot (black = 1.0) of the vector **g** laid out spatially to correspond to the input field. Each point represents the dg_i associated with one translational mapping of a subfield of the input field r_{fwd} to b_{fwd}. The location of the point is the position in the input field of the lower left corner of the mapped subfield. Successive frames in (a) coincide with successive peaks of oscillatory activity (see Chapter 5). Corresponding frames of (b) represent approximately equal intervals relative to the full interval of convergence.

In the comparison test both circuits' memories are trained with eight patterns, one of which is the target memory, Fig. 2-3(e), and seven of which are non-tiled distractors. The distractors consist of 40 x 40 sections taken from the original edge-detected image lying partially or entirely outside the test image area. A rotation, an inversion and a partial displacement of the target memory pattern are also used as distractors. Fig. 2-3(c) is used as the input test pattern. (The location of a target along the edge demonstrates that the initial centralized distribution of the b_{fwd} signal to memory does not interfere with finding matches near the boundaries of the input pattern.)

Fig. 2-4(a) shows spatial density plot of r-match cell activity (equivalent to **g**) during convergence of neuronal circuit. Fig. 2-4(b) shows spatial density plot of **g** values during convergence of algorithmic circuit. Each sequence is extracted at corresponding intervals for both simulations.[6]

Though the target area of the input and target memory in the test shown above are exact matches, that need not be the case, as will be seen below. The algorithmic circuit simulation is both far more compact and faster than the neuronal, allowing far larger and more demanding tests to be run. The single layer algorithmic simulation has been successfully run with mappings corresponding to 36 rotations (at 10-degree intervals) for every translation: 90,000 mappings in the largest test. With such circuits target memory patterns which are rotated with respect to the test image, such as Fig. 2-3(f), are readily recognized.

For tests with large numbers of active memories, additional distractor memories, an example of which is shown in Fig. 2-3(g), are assembled from random permutations of tiles selected from the target memory and regular distractors. These are not intended as realistic memory patterns, but rather provide controlled input conditions which analysis suggests should stress the circuit. As will be seen later in this chapter large numbers of initial components in the superpositions increase the probability of failure. The tests that employ large numbers of tiled distractors emulate a condition in which memory is populated with many images having high correlations between them. This is a more realistic condition than deliberately randomized images because it represents the task of recognition within a class of images whose members share some subset of features and are distinguished by other features. Under most variants of the test correct convergence is achieved without difficulty. However, certain combinations of input pattern, memory patterns and deployed mappings can create an erroneous convergence condition (the *collusion* condition mentioned earlier). This is discussed at length later in this chapter.

2.4 Multilayer Circuits

While the map-seeking dynamic is capable of handling very large numbers of mappings it should be evident that combinatorial explosion precludes accommodating all translations, rotations, scalings and other necessary

---

[6] Note: the map reference point origin in these tests is lower left hand corner of memory pattern. Absence of activity in top section of each **g** plot results from subthreshold "edge condition" of mappings along top section. Mappings in lower right wrap to left edge of input image. These edge conditions in algorithmic circuit apply only to this test, to mimic neuronal circuit as closely as possible.

transformations by creating a separate mapping for each possible combination. Fortunately the inherent characteristics of the dynamic allow independent sets of mappings to be accommodated in separate layers of circuitry. For example, the 90,000 mappings necessary in a single layer circuit to implement 50x50 translations with 36 possible rotations for each translation, can be reduced to 2536 mappings in a two-layer circuit.

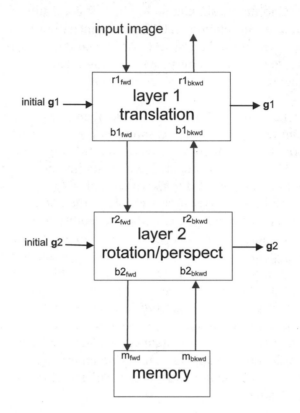

Fig. 2-5: Two-layer circuit with memory.

A two-layer circuit can be set up with translation mappings in layer 1 and rotation mappings in layer 2. Thus, instead of requiring maps for the product of the number of each kind of transformation, only the sum is required. The input image is fed to r_{fwd} of layer 1. Layer 1 presents a translational superposition to r_{fwd} of layer 2. Layer 2 presents the combined translational-rotational superposition to memory, and similarly for the path backward. Competition between the mappings takes place within each layer, converging to a single mapping for translation in layer 1 and a single mapping for rotation in layer 2.

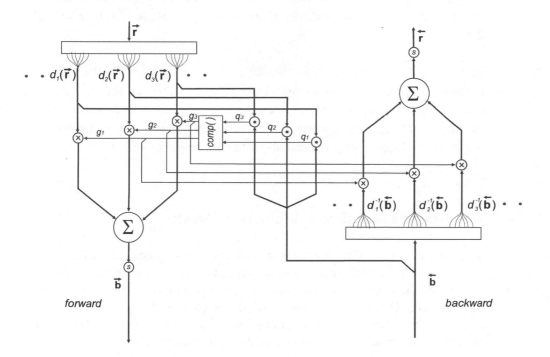

Fig. 2-6: Dataflow within one layer for multilayer circuit. Signal paths are vector (indicated with a heavy line and bold symbol) or scalar (indicated by light line and italic symbol). Each bundle of scalar paths emerging from the rectangles designates a mapping $d_i(\vec{\mathbf{r}})$ or the associated inverse mapping $d_i^{-1}(\tilde{\mathbf{b}})$.

Each layer is architecturally identical, as shown in Fig. 2-6. The r_{bkwd} path, not present in the one-layer circuit, allows two or more of these layers to be cascaded, as shown in Fig. 2-5.

The signal on the r_{bkwd} path is the superposition of inverse mappings $d_i^{-1}(\tilde{\mathbf{b}})$ of the signal on the b_{bkwd} path, each attenuated by the associated coefficient g_i. Again ignoring scaling,

$$\tilde{\mathbf{r}} = \sum g_i \cdot d_i^{-1}(\tilde{\mathbf{b}})$$

The b_{fwd} path of layer 1 (designated $b1_{fwd}$) feeds the r_{fwd} path of layer 2 (designated $r2_{fwd}$) and the r_{bkwd} path of layer 2 ($r2_{bkwd}$) feeds the $b1_{bkwd}$ path of layer 1 ($r1_{fwd}$). More than two layers may be cascaded this way. The last layer is connected to the memory stage via b_{fwd} and b_{bkwd} as shown in Fig. 2-5.

The principles of operation for a two-layer circuit are essentially the same as those already discussed for a single-layer circuit. However, the forward inputs to layer 2 are not static as they were in the single-layer case. As a result the convergence proceeds more quickly in layer 1, and as it resolves, the inputs to layer 2 stabilize and allow the convergence in layer 2 to accelerate.

Resolving the Mappings in Multilayer Circuits

The progression of resolving mappings during convergence can be best seen by monitoring the signals crossing between the two layers. A clear demonstration of these signals is produced when using the circuit in a shape-from-motion task. The task is to determine the angles of the two visible faces relative to the observer by determining the changes in perspectivity between the two views. That aspect of the demonstration will be discussed in Chapter 3. It is presented in this chapter to illustrate the state of the signal superpositions as the circuit converges to a recognition state.

The input images are two views of the 3D cube seen in Fig. 2-7(b,c) from the viewpoints shown in Fig. 2-7(a). The two views differ slightly due to the displacement of the viewpoints. Both views are converted to circuit input data by edge filtering. The data from the first view, Fig. 2-8(a), is captured in memory. The data from the second view is used twice, by deleting part of the image to simulate a fixation on each visible face, Fig. 2-8(b,c).

The circuit used here has 2500 translational mappings in layer 1, without the boundary overlaps seen in the previous demonstration. In this, as in all subsequent algorithmic circuit demonstrations, graphic representations of **g** locate the origin of the mapping in the middle of the subfield instead of its corner. Layer 2 has 110 mappings: 10 rotations (at 10° intervals) times 11 foreshortenings in the y-axis of the rotated field. (Perspective mappings will be addressed further in Chapter 3 and Chapter 7.) In this demonstration the capture view, Fig. 2-8(a) is loaded into a memory, then the two "fixation" views, Fig. 2-8(b and c) are presented to the circuit in sequence and a convergence is reached for each, as can be seen in the layer 1 and layer 2 **g** data in Fig. 2-8(d-h).

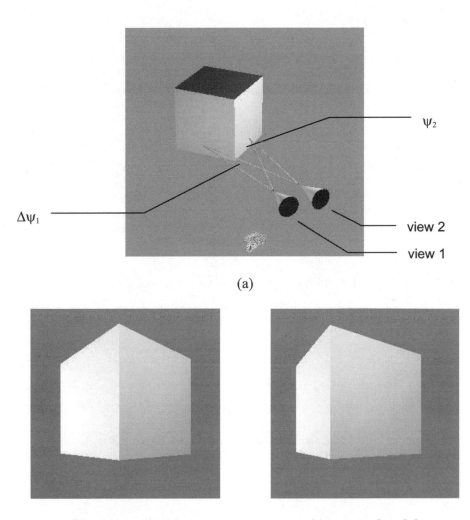

(a)

(b) capture: view 1   (c) target: view 2,3

Fig. 2-7: Geometry of test case. $\Delta\psi_1$, $\Delta\psi_2$ = -13, -14 degree (right rotation view 1 to view 2, left face, right face). ψ_1, ψ_2 = 30, -36 degrees (angle between view 1 and left face, right face).

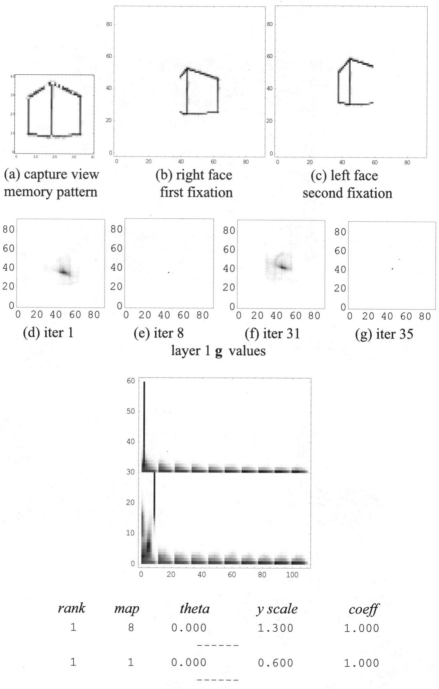

(a) capture view
memory pattern

(b) right face
first fixation

(c) left face
second fixation

(d) iter 1   (e) iter 8   (f) iter 31   (g) iter 35

layer 1 **g** values

rank	map	theta	y scale	coeff
1	8	0.000	1.300	1.000

1	1	0.000	0.600	1.000

(h) layer 2 **g** values

Fig. 2-8: Two-layer circuit input and data. (a - c) capture and target data; (d - g) layer 1 mappings; (h) upper: Density plot of layer 2 **g**. Iterations are enumerated along vertical axis, indices *i* of **g** enumerated along horizontal axis. Change of input at iter 31. (h) lower: Tabular form of final non-zero layer 2 **g**. *rank* is the order in descending *coeff* value of surviving mappings (represented by non-zero **g** values), *map* is index of g_{idx}. For clarity, *map* is translated into the parameters *theta* and *y scale* which characterize the geometry (rotation and foreshortening) of the mapping corresponding to g_{idx}. *coeff* is the final value of g_{idx}.

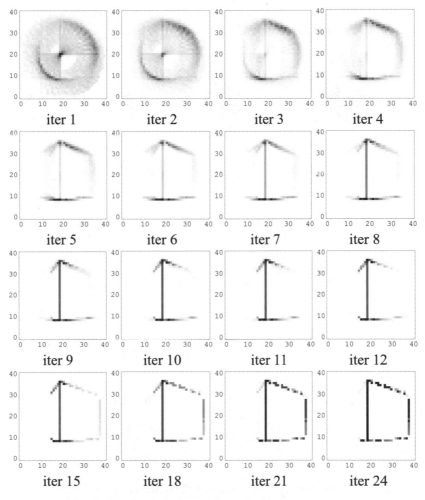

Fig. 2-9: $r2_{fwd}$ intersect $r2_{bkwd}$ during convergence.

The series of panels in Fig. 2-9 shows the progression of pairwise intersections of signals on $r2_{fwd}$ and $r2_{bkwd}$ during the first phase of convergence (the right face of the cube). Intersection is implemented as $\bar{r} \cdot \bar{r}$. This is a compact way of revealing which signals are causing matching in both layers because $\bar{r}$ carries surviving forward mappings (translations) of the input in layer 1 and $\bar{r}$ carries surviving backward mappings (rotations and y-axis foreshortenings) of memory in layer 2.

The data in Fig. 2-10 shows the progression of the superposition on $r1_{bkwd}$. Since this is the endpoint of the backward pathways it contains both the superpositions of the layer 2 mappings (rotations and y-axis foreshortenings) and the layer 1 mappings (translations) on the pattern in memory. Since there is only one trained memory in this demonstration the effect of the combined mappings is evident.

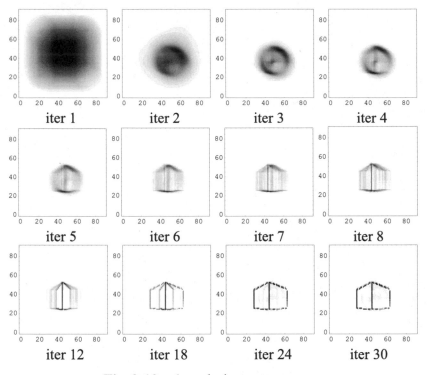

Fig. 2-10: $r1_{bkwd}$ during convergence.

2.5 Other Multilayer Circuit Demonstrations

The circuit application just discussed is designed to determine the orientation of the cube surfaces and is therefore an "interpretation" task rather than a "recognition" task. Multilayer circuits are capable of matching target patterns to one of a number of memory patterns using composed mappings through two, three or more layers.

(a) aerial photo

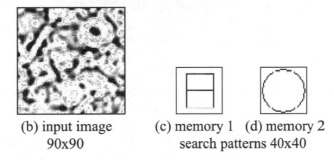

| (b) input image | (c) memory 1 | (d) memory 2 |
| 90x90 | search patterns 40x40 |

Fig. 2-11: Aerial photo test data.

To demonstrate this capability a two-layer and an equivalent three-layer circuit are set up to search a World War 2 aerial reconnaissance photo, Fig. 2-11(a), for typical building top views. Both circuits have 2500 translational mappings in layer 1. The two-layer has 1573 combined scaling and rotation mappings in layer 2 and the three-layer has 13 rotational mappings in layer 2 and 121 scaling mappings in layer 3.[7] The three-layer circuit converges almost identically to the two-layer, as seen in Appendix H. The aerial photo was processed by increasing contrast, followed by low pass filtering and naive edge filtering. The resulting input data, Fig. 2-11(b), is both noisy and fragmented. The search patterns, Fig. 2-11(c,d) are synthetic, and differ from the targets in scale, proportion and orientation.

It is worth noting that to find each target by simple correlation would require nearly 8,000,000 dot-products. By exploiting the superposition ordering

---

[7] These comprise rotations from -60° to +60° by 10° increments, and scalings in the memory x and y axes from 0.75 to 1.25 by 0.05 increments. For the circle pattern only the range from 0.75 to 1.05 is effective since the dimensions of $r2_{fwd}$ and $r2_{bkwd}$ are the same as $b2_{fwd}$ and $b2_{bkwd}$ in this circuit.

property and skipping the mappings with subthreshold coefficients after the first iteration, an average of 30,000 dot-products per recognition are required.

This algorithmic circuit used in this test differs from those already demonstrated in that it implements competition among memories as well as among mappings. The memory competition is not initially active, but commences several iterations after the start of recognition. The purpose is to completely suppress the response of the non-winning memory or memories. Since there are multiple targets in the input image which are potential matches to each search pattern the circuit implements sequential attention shifts every 25 iterations to find all possible matches. This is accomplished by zeroing areas in the input field which are successfully matched, and resetting all the mapping coefficients to the start condition. Four attention shifts take place in the 100 iterations of this test. Targets are matched in the first three, and on the third no match survives, indicating a *non-recognition* state.

The behavior of the circuit is illustrated in Fig. 2-12. Density plots of the intersection of layer 1 forward and backward signals ($rl_{fwd} \cap rl_{bkwd}$) indicate where the matched pattern is located in the input field. Layer 2 mappings at the end of each 25 iteration attention "span" are sorted by coefficient **g** and the geometric parameters for each of the best few mappings are listed. Data for the four attention shift sequences is shown. The orientation, x-scale and y-scale parameters for the best mappings identify the transformation between the idealized search pattern and the located targets in the input image.

1. The dome outline located in the first span is somewhat elliptical and inclined 20-30 degrees to the left (iter 25). This sequence shows evidence of the beginning of a collusion which responds to the partly rectangular pattern to the lower left of the dome in iteration 5. The collusion has broken up completely by iteration 10.

2. The small squarish building located in the second span is by coincidence similarly scaled and oriented though the proportion of the rectangular top view is very different (iter 50). At the end of convergence two mappings with similar parameters have survived with nearly the same coefficient: both mappings have resulted in nearly equally good matches of this target pattern as a consequence of the blur of the input data. In cases like this one interpolation between parameters of the best surviving mappings yields meaningful data about the target, but that may not be true when different portions of the target pattern are matched by the surviving mappings.

3. The elongated diagonal structure in the upper left corner of the input is matched to the upper half of the rectangular search pattern compressed and rotated left about 40 degrees (iter 75).

4. The part of the input pattern initially matched in the fourth span fails to constitute a sufficient match, and falls below threshold soon after the tenth iteration of the span. As a result no mappings survive in either layer (iter 100).

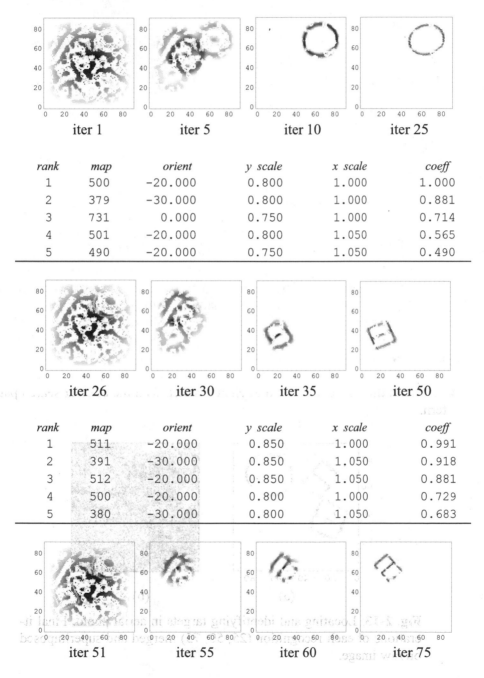

iter 1   iter 5   iter 10   iter 25

rank	map	orient	y scale	x scale	coeff
1	500	-20.000	0.800	1.000	1.000
2	379	-30.000	0.800	1.000	0.881
3	731	0.000	0.750	1.000	0.714
4	501	-20.000	0.800	1.050	0.565
5	490	-20.000	0.750	1.050	0.490

iter 26   iter 30   iter 35   iter 50

rank	map	orient	y scale	x scale	coeff
1	511	-20.000	0.850	1.000	0.991
2	391	-30.000	0.850	1.050	0.918
3	512	-20.000	0.850	1.050	0.881
4	500	-20.000	0.800	1.000	0.729
5	380	-30.000	0.800	1.050	0.683

iter 51   iter 55   iter 60   iter 75

rank	map	orient	y scale	x scale	coeff
1	265	-40.000	0.850	0.800	1.000
2	254	-40.000	0.800	0.800	0.813
3	144	-50.000	0.850	0.800	0.560
4	243	-40.000	0.750	0.800	0.505
5	276	-40.000	0.900	0.800	0.405

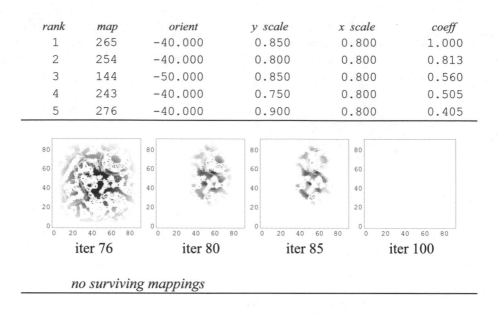

iter 76   iter 80   iter 85   iter 100

no surviving mappings

Fig. 2-12: Attention shifts in a single input image, aerial photo.

The layer 1 intersection patterns from the end of each successful recognition are merged in Fig. 2-13(a) and superimposed on the raw input image in Fig. 2-13(b). At least one obvious target within the input field was missed: the smaller dome at the bottom of the field. An inspection of the edge filtered input data, Fig. 2-11(b), reveals that only the right half of the dome had a sufficient gradient at the boundary to be detected by the naïve edge filtering used, and thus the dome failed to elicit a match with the circular search pattern.

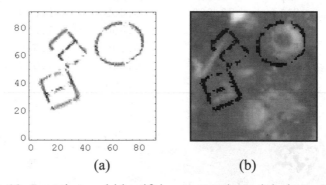

(a)   (b)

Fig. 2-13: Locating and identifying targets in aerial photo. Final iterations of each recognition (25, 50, 75) merged and superimposed on raw image.

Manual analysis of the raw photo reveals that a boundary gradient does in fact exist in the shadow of the dome, but the local contrast has to be raised significantly to produce a response from the edge filter. This suggests that the $rl_{fwd} \cap rl_{bkwd}$ signal could be productively used as an adaptive local contrast gain control signal to processing stages ahead of layer 1.[8]

The responses of the two memories in the last iteration of each recognition are shown in Table 2-1. The effect of the memory competition function is evident in the zero m_{fwd} value for the losing memory in each span.

iter	$m_{fwd}\ 1$	$m_{fwd}\ 2$
25	0.000	0.620
50	0.571	0.000
75	0.498	0.000
100	0.000	0.000

Table 2-1: Memory response in final iteration of each recognition

In the absence of memory competition the losing memory would show some response to the incidental matches and noise in the input field. The same test has been performed with several distractor memories in addition to the search patterns Fig. 2-11(c,d). Memory competition also suppresses spurious responses among the distractor memories.

It has been noted earlier in this chapter that all mappings in these tests (and all others in the book) were implemented as one-to-one projections, without interpolation of any sort. However, input image blur compensates for quantization and improves the overlap with the memorized templates. It may be that biological visual systems exploit eye jitter similarly: as a parsimonious form of interpolation. Interpolation can be incorporated in the mappings (this has been confirmed experimentally), but it is far less resource intensive to blur the input.

Practical Limits

The simple filtering applied to the aerial photo above produces input data that is near the limit of effective processing by the circuit used. Blur results in a high percentage of non-zero pixels: in other words low sparsity. Because the blur cross section remains constant in the input image, the shapes of smaller targets are less well defined and the selectivity of matching for a

---

[8] This echoes the suggestion by many neuroscientists that backward projections from higher cortical areas may provide gain control signals to V1 and the LGN.

specific shape diminishes. This imposes a lower bound on the size of target that can be searched for reliably. Fig. 2-14 illustrates the same test as seen in the previous section, but with a range of x and y scalings about half those used previously. The circuit finds a match to the circle pattern which does not correspond to any structure visible by eye in the photo. Fig. 2-14(c), in which the located circle has been subtracted by an attention shift, reveals that the circuit appears to be matching a combination of negative space (a shape defined by boundaries of surrounding structures) and noise.

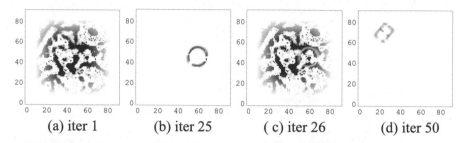

| (a) iter 1 | (b) iter 25 | (c) iter 26 | (d) iter 50 |

Fig. 2-14: Effect of low image sparsity. Range of mapping x and y scale 0.4 to 0.75 results in spurious match for circle pattern. Rectangular pattern match is valid.

The immediate conclusion is that the blur cross section simply limits how small a pattern may be usefully sought. Sharper boundaries, and therefore sparser input, can be achieved by more sophisticated edge filtering, but with raw data of this quality that results in meandering edges unless some a priori assumptions are made about the nature of the edges likely to occur. A more robust approach to increasing sparsity is to probabilistically encode the local orientation of edges while leaving the spatial resolution the same. (See Appendix B.) Alternatively the simple dot product used in the matching algorithm can be replaced with more sophisticated matching algorithms that require the spread of edge pixels in the input image to "track" the mapped search pattern over some significant distance. These approaches to refining a circuit's criterion for a match are matters of evolutionary or human engineering, and are outside the scope of this book. The map-seeking principle which exploits them remains the same.

2.6 Convergence and the Ordering Property

The ordering property is responsible for directing the convergence to select the proper memory and mapping. As stated early in this chapter, when matches are computed between a superposition and a set of patterns the degree of match will with high probability be the greatest for that pattern which has the closest match to one of the components of the superposition. That probability is dependent on a number of factors, including the number of components in the superposition and the sparsity of the encoding. The degree of match between a solo pattern and a superposition of patterns is a function of the sum of "interior" matches between the solo pattern and each of the component patterns of the superposition. Therefore it is necessary to examine the "interior" matches. We will work from the simpler to the more complex situations that may be encountered by a circuit.

In the following analysis the scaling functions $s(\ )$ are ignored since they are applied to the entire superposition in each direction therefore do not affect the ordering within each superposition.

The case of two close competitors in the input field

Suppose the input field contains two subfields which when mapped, $d_x(\vec{r})$ and $d_y(\vec{r})$, corresponding strongly to two memories m_x and m_y with weights respectively $\mathbf{w}_x$ and $\mathbf{w}_y$, and all the other subfields have very weak or no response from memory. Assume the response of m_x to $d_x(\vec{r})$ exceeds, if only by a little, the response of m_y to $d_y(\vec{r})$. Starting with $g_x = g_y = 1.0$

$$\mathbf{w}_x \bullet g_x \cdot d_x(\vec{r}) > \mathbf{w}_y \bullet g_y \cdot d_y(\vec{r}) \qquad \text{eq. 2-7}$$

Since both memories are strongly responding the angle between $d_x(\vec{r})$ and $\mathbf{w}_x$ is equal or close to the angle between $d_y(\vec{r})$ and $\mathbf{w}_x$,

$$\mathbf{w}_y \bullet g_x \cdot d_x(\vec{r}) \approx \mathbf{w}_x \bullet g_y \cdot d_y(\vec{r}) \qquad \text{eq. 2-8}$$

Therefore, with high probability

$$\mathbf{w}_x \bullet \left(g_x \cdot d_x(\vec{r}) + g_y \cdot d_y(\vec{r})\right) > \mathbf{w}_y \bullet \left(g_x \cdot d_x(\vec{r}) + g_y \cdot d_y(\vec{r})\right) \quad \text{eq. 2-9}$$

Let

$$u = f\left(\mathbf{w}_x \bullet \left(g_x \cdot d_x(\vec{r}) + g_y \cdot d_y(\vec{r})\right)\right) \qquad \text{eq. 2-10}$$

$$v = f\left(\mathbf{w}_y \bullet \left(g_x \cdot d_x(\vec{r}) + g_y \cdot d_y(\vec{r})\right)\right) \qquad \text{eq. 2-11}$$

Note that the arguments of $f(\ )$ in eq. 2-10 and eq. 2-11 appear in the inequality of eq. 2-9, and $f(\ )$ is monotonic so

$$u > v \qquad \text{eq. 2-12}$$

Substituting in eq. 2-5

$$q_x' = \bar{\mathbf{b}} \bullet d_x(\bar{\mathbf{r}}) = (u \cdot \mathbf{w}_x + v \cdot \mathbf{w}_y) \bullet d_x(\bar{\mathbf{r}}) \qquad \text{eq. 2-13}$$

$$q_y' = \bar{\mathbf{b}} \bullet d_y(\bar{\mathbf{r}}) = (u \cdot \mathbf{w}_x + v \cdot \mathbf{w}_y) \bullet d_y(\bar{\mathbf{r}}) \qquad \text{eq. 2-14}$$

Since $u > v$

$$q_x' > q_y' \qquad \text{eq. 2-15}$$

and *comp( )* is monotonic and amplifies differences

$$g_x' \gg g_y' \qquad \text{eq. 2-16}$$

("$\gg$" here meaning *more greater-than* than "$>$")

Hence when these two coefficients g_x' and g_y' are applied to the next iteration, the inequality in eq. 2-7 is increased in favor of $\mathbf{w}_x$. Any inequality in eq. 2-8 is increased in favor of $\mathbf{w}_y$ but by a much smaller amount. So the inequality in eq. 2-9 increases in favor of $\mathbf{w}_x$, and this passes on to further increase the inequality in eq. 2-16.

More complex competitions

The situation described above is the basis for establishing by induction a partial ordering of many components. This is applicable to circumstances of two or a few competitors with good matches to distinct memories, or to collusion situations. However, with more than two components, many small terms must be considered. The magnitude of those terms depends on the case.

Expanding eq. 2-5

$$q_i' = \bar{\mathbf{b}} \bullet d_i(\bar{\mathbf{r}}) = \left(\sum_k f\left(\left(\sum_j g_j \cdot d_j(\bar{\mathbf{r}})\right) \bullet \mathbf{w}_k\right) \cdot \mathbf{w}_k \right) \bullet d_i(\bar{\mathbf{r}}) \qquad \text{eq. 2-17}$$

Define

$$BB = \sum_k f\left(\left(\sum_j g_j \cdot d_j(\bar{\mathbf{r}})\right) \bullet \mathbf{w}_k\right) \cdot \mathbf{w}_k \qquad \text{eq. 2-18}$$

Again consider two subfields $d_x(\bar{\mathbf{r}})$ and $d_y(\bar{\mathbf{r}})$. These correspond to memories with weights $\mathbf{w}_x$ and $\mathbf{w}_y$. The first of these will be considered the exact or strong match. The strength of the second match will depend on application.

To simplify the notation we will assume here that $f(z) = z$. (This is the worst case assumption since the non-linear $f(\ )$ functions used in practice reduce the contribution of smaller arguments more effectively.) This assumption allows us to partition BB into nine terms.

$BB=$

$c1$: $P_{(x,x)}$ +

$c2$: $P_{(y,y)}$ +

$c3$: $P_{(x,y)}$ +

$c4$: $P_{(y,x)}$ +

$c5$: $P_{(x,\,\overline{x \cup y})}$ +

$c6$: $P_{(y,\,\overline{x \cup y})}$ +

$c7$: $P_{(\overline{x \cup y},\,x)}$ +

$c8$: $P_{(\overline{x \cup y},\,y)}$ +

$c9$: $P_{(\overline{x \cup y},\,\overline{x \cup y})}$ eq. 2-19

where

$$P_{(j\_range,\,k\_range)} = \sum_{k\_range} \left(\left(\sum_{j\_range} g_j \cdot d_j(\vec{\mathbf{r}}) \right) \bullet \mathbf{w}_k \right) \cdot \mathbf{w}_k$$

and

$\overline{x \cup y}$ means all elements excluding x and y.

To determine the ordering for q_x' and q_y'

$$q_x' = BB \bullet d_x(\vec{\mathbf{r}})$$ eq. 2-20

$$q_y' = BB \bullet d_y(\vec{\mathbf{r}})$$ eq. 2-21

we need to examine all the partitions of BB. We already know the relationship of some of the terms of BB from the previous case, eq. 2-7 to eq. 2-15.

$$(c1+c2+c3+c4) \bullet d_x(\vec{\mathbf{r}}) > (c1+c2+c3+c4) \bullet d_y(\vec{\mathbf{r}})$$

Therefore the other terms of *BB* determine whether that ordering is preserved or reversed. Recall that the j subscript denotes the range of input subfields affecting the partition and the k subscript denotes the range of memories affecting the partition

If in the remainder of memories, $\mathbf{w}_k$, $k = x \cup y$, there is no bias toward matching either $d_x(\vec{\mathbf{r}})$ or $d_y(\vec{\mathbf{r}})$ then

$$(c5 + c6) \bullet d_x(\vec{\mathbf{r}}) \approx (c5 + c6) \bullet d_y(\vec{\mathbf{r}})$$

If in the remainder of the input subfields, $d_j(\vec{\mathbf{r}})$, $j = x \cup y$, there is no bias toward supporting either $\mathbf{w}_x$ or $\mathbf{w}_y$, then

$$(c7 + c8) \bullet d_x(\vec{\mathbf{r}}) \approx (c7 + c8) \bullet d_y(\vec{\mathbf{r}})$$

So if the small terms are not biased in aggregate the ordering is preserved in the presence of many minor fields since $c9 : P_{\overline{(x \cup y, \, x \cup y)}}$ does not affect the situation, at least to the first order. (Though it might to a second order, through affecting participants in partitions $c7$ and $c8$.)

$$c9 \bullet d_x(\vec{\mathbf{r}}) \approx c9 \bullet d_y(\vec{\mathbf{r}})$$

Partitions c7, c8 and c9 are pruned first in the convergence.

The probability and magnitude of any biases depends on the statistics of the various $d_j(\vec{\mathbf{r}}) \bullet \mathbf{w}_k$, as listed in the partitions above. This will be analyzed below.

In the case of collusion we assume no significant matches between any memory and any subfield other than $d_x(\vec{\mathbf{r}})$ and $\mathbf{w}_x$. A somewhat different partition is necessary because a set of memories $\mathbf{w}_{y_i}$ where $y_i \in Y$ support the match with input subfield $d_y(\vec{\mathbf{r}})$. Now

$BB=$

$c1$:	$P_{(x,x)}$	$+$
$c2$:	$P_{(y,Y)}$	$+$
$c3$:	$P_{(x,Y)}$	$+$
$c4$:	$P_{(y,x)}$	$+$

$c5$:　　$P_{(x,\,\overline{x \cup Y})}$ +

$c6$:　　$P_{(y,\,\overline{x \cup Y})}$ +

$c7$:　　$P_{(\overline{x \cup y},\,x)}$ +

$c8$:　　$P_{(\overline{x \cup y},\,Y)}$ +

$c9$:　　$P_{(\overline{x \cup y},\,\overline{x \cup Y})}$　　　　　　　　　　　eq. 2-22

$c1$: $P_{(x,x)}$ is the component of primary memory response to target $d_x(\vec{r})$.

$c2$: $P_{(y,Y)}$ is the component of collusion memories response to $d_y(\vec{r})$.

$c3$: $P_{(x,Y)}$ is the component of collusion memories response to $d_x(\vec{r})$.

$c4$: $P_{(y,x)}$ is the component of primary memory response to $d_y(\vec{r})$.

$c5$: $P_{(x,\,\overline{x \cup Y})}$ is the component of incidental memory response to $d_x(\vec{r})$.

$c6$: $P_{(y,\,\overline{x \cup Y})}$ is the component of incidental memory response to $d_y(\vec{r})$.

$c7$: $P_{(\overline{x \cup y},\,x)}$ is the component of primary memory response to background.

$c8$: $P_{(\overline{x \cup y},\,Y)}$ is the component of collusion memories response to background.

$c9$: $P_{(\overline{x \cup y},\,\overline{x \cup Y})}$ is the component of incidental memory response to background.

Partitions c7, c8 and c9 are eliminated early in the convergence. Partitions c5 and c6 (assuming no memory competition mechanism) are small and unless highly biased to support one or the other input fields do not contribute to one side or the other. By assumption c4 is negligible.

Therefore the outcome of the collusion depends on

$$(c1 + c2 + c3) \bullet d_x(\vec{r}) \ ? \ (c1 + c2 + c3) \bullet d_y(\vec{r})$$

By assumption the cross responses are small, so effectively

$$c2 \bullet d_x(\vec{r}) \approx c1 \bullet d_y(\vec{r})$$

Therefore the survival of the collusion is dominated by

$$(c1 + c3) \bullet d_x(\vec{r}) \ ? \ (c2 + c3) \bullet d_y(\vec{r})$$

over the course of the convergence. Assuming no memory competition the primary target/memory match will succeed only if

$$c1 \bullet d_x(\vec{r}) > c2 \bullet d_y(\vec{r}) + \left(c3 \bullet d_y(\vec{r}) - c3 \bullet d_x(\vec{r})\right)$$

If c3 has any net effect it will be to support the collusion.

However, if memory competition is implemented then the primary target/memory match can succeed even if initially

$$c1 \bullet d_x(\vec{r}) \leq c2 \bullet d_y(\vec{r}) + \left(c3 \bullet d_y(\vec{r}) - c3 \bullet d_x(\vec{r})\right)$$

because the pruning of the contributors to c2 and c3 can reverse that relationship in the course of convergence.

Probability of satisfying ordering conditions

What factors affect the various "minor" conditions which must be satisfied to preserve the ordering? More precisely, what are the expected values and distribution of values of $d_j(\vec{r}) \bullet \mathbf{w}_k$, as enumerated in the partitions? This of course depends on the nature of the images and how they are encoded, and that is too large a space to analyze in depth here. The major factors are the distribution of potentially matching elements in the set of patterns contributing to the superpositions and the sparsity with which the patterns are encoded. The discussion below presents a method of analysis, using a simple pattern generating function with well known properties.

To get a sense of the probabilities we can use a 2D binomial random walk as the contour generating process and estimate the probability function of the degree of match between two 2D binomial random walks. Considering only locations *(x,y)* where $x + y = n$, and the probability of taking an *x* step or a *y* step are both 0.5, the probability of landing on a given *(x,y)* on the *n*-th step is approximately

$$P_n(x,y) = C(n, \frac{n}{2}) \cdot 0.5^n$$

For walks of n=1..200, the average value of P_n is about 0.08. The probability of *k* hits in a walk of length *m* is computed

$$P(m,k) = C(m,k) \cdot P_{n\_avg}{}^k \cdot (1 - P_{n\_avg})^{m-k}$$

The distribution of probability of hits for a walk of m=200, $P_{n\_avg} = 0.08$ is shown in Fig. 2-15. The expected value of hits is 16.

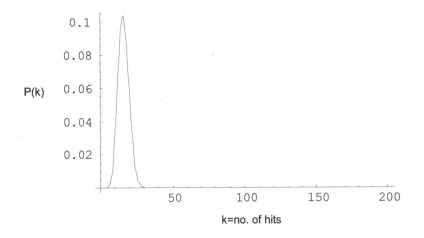

P(k)

k=no. of hits

Fig. 2-15: Probability of hits in binomial walk of m=200.

Therefore 16/200=0.08 can be taken as the expected value of $d_j(\vec{r}) \bullet \mathbf{w}_k$ for 200 pixel pattern contours. This assumes optimal translation to start the two from a common origin. The conditions that need to be met, from above, are $c5 \approx c6$ and $c7 \approx c8$. Given the nature of the distribution shown in Fig. 2-15 large numbers of contributors would be needed to effect a significant inequality in those conditions. Because of the low expected value, small variance and symmetry the conditions for maintaining the ordering are met with high probability, with this generating process at least. For longer walks the normalized expected value falls and the normalized variance decreases. (In the demonstrations patterns typically have 100-300 contour pixels.)

To use the same approach to compute probabilities for collusions multiple shorter walks are used. In this case we are interested in the probability of finding a number of components that will collude to match most of a full pattern. For walks of length 10, P_n is about 0.3 and the probability of a 70% or better match is about 0.01. The probability of a 70% or better match over a pattern of total length 100 by a collusion of 10 walks (even assuming the correct translation for each) is about 10^{-20}, so even with a large number of candidates from which to pick colluding components such an assembly is extremely unlikely.

Practical experience with pixel encoded images indicates that at least temporary inversions in superposition orders are far more common than the analysis above would suggest. That is because the random walk is an unrealistic generating process for pixel encoded images. Statistics of natural images show high probability of short straight lines and in images of man made environments long straight lines predominate. And because jitter, blur or interpolation is necessary to make matching reliable, contours are effectively more than a pixel wide. But while the random walk is not a realistic generating process for pixel encoded images, such as those which appear throughout this book, it has characteristics that approximate those of more sophisticated and biologically realistic contour encodings. As will be discussed in later chapters, endstops or inflections most likely have distinct encodings which weigh heavily in the match. Also straight lines are likely encoded by segments, possibly of a variety of lengths. In the limit this sort of encoding completely discounts the correlation of straight sections and becomes a walk of uncorrelated steps. Here the probability of a step in a particular direction and particular length is $1/DL$, where D is the number of direction codes and L the number of length codes (with the endstop encoded as zero length, for example).

With richer encodings the number of steps, m, needed to describe an image contour decreases, but so does the probability of making the right step or sequence of steps, P_{n_avg}, to effect a hit. Fig. 2-16 shows the distribution of hits for a walk of length 50 when $P_{n\_avg} = 0.02$. Normalized expected value for $d_j(\vec{r}) \bullet \mathbf{w}_k$ is about 0.03. As DL increases P_{n_avg} decreases and the likelihood of a hit and the variance both decrease.

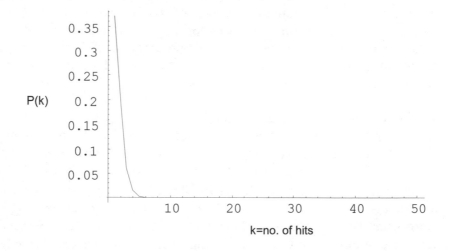

Fig. 2-16: Probability of hits in binomial walk of m=50.

This discussion points to an important criterion for image encoding schemes: good ones produce small expected values, small variance and symmetry in the distribution of values of $d_j(\vec{r}) \bullet \mathbf{w}_k$ for incidental matches.

The preceding discussion applies to collusions created by multiple memory patterns. Collusions created by multiple mappings would have a similar character if the implemented mappings were somewhat random and if the competition in the circuit were exclusively (or dominantly) between memories instead of mappings. However, since the mappings in practice are fairly close samplings of a continuum, and even in circuits with memory competition the mapping competition is dominant, the competitive situation faced by a collusion of partial mappings is quite different. Partial mappings by definition will lose the competition with a more complete mapping, and the near continuum of implemented mappings makes it highly probable that a complete or nearly complete mapping will exist if there is a memory pattern which is a meaningful transform of a target pattern. Terrain interpretation is the situation in which mapping collusions would be seen and they do not occur for the reason just given.

Factors affecting the probability of correct convergence

It will now be apparent that a number of factors affect the reliability of convergence. As mentioned earlier, the superposition signals must not saturate the channels that carry them, for saturation erases the potentially small differences which distinguish the ordering when there are large numbers of superposition components. And similarly, the monotonicity (if not linearity) of the channel must be adequate to reliably preserve those small differences, at least statistically. The magnitude of those differences is itself dependent on the number of contributing patterns and their sparsity, and to some degree the nature of the patterns themselves. These are technical details and do not affect the overall purpose of this book. They are discussed further in Chapter 5 and Appendix A. All of the factors in convergence are influenced by the sparsity of the encoding of the input patterns. This is discussed in Appendix B.

2.7 Mapping Modes: Seeking, Driven, Restricted

As will be seen in later chapters circuits may be combined to execute a sequence of operations necessary to solve complex visual and motor tasks. Not all of the circuits in these combinations operate in the same mode: some solve inverse mapping problems, others transform data on the basis of externally specified parameters. Three operating *modes* of the circuits will be used to characterize them throughout the book: *map-seeking, driven-mapping* and *restricted-mapping*.

To define these a bit of mathematical looking notation (which will recur throughout the book) is used: a shorthand for the mapping process that allows a terse description of what a mapping circuit does, with no attention to how it does it.

$$T_x(\mathbf{y}) = \mathbf{z}$$
eq. 2-23

simply means that transformation T_x applied to data pattern y produces data pattern z. If T_x is a 135° right rotation, and y is an triangle with an apex pointing north, then z is the same triangle twirled so that apex is pointing southwest . The inverse of T_x is denoted T_x^{-1}, so

$$T_x^{-1}(\mathbf{z}) = \mathbf{y}$$
eq. 2-24

simply meaning that T_x^{-1}, the inverse of T_x , applied to data pattern z produces data pattern y. T_x^{-1} in this instance is a 135° left rotation.

The transformations T_x and T_x^{-1} may be composed of multiple transformations

$$T_x(\mathbf{y}) = t_{x1}\big(t_{x2}\big(t_{x3}(\mathbf{y})\big)\big) \quad \text{and} \quad T_x^{-1}(\mathbf{z}) = t_{x3}^{-1}\big(t_{x2}^{-1}\big(t_{x1}^{-1}(\mathbf{z})\big)\big) \quad \text{eq. 2-25}$$

or more gracefully

$$T_x(\mathbf{y}) = t_{x1} \circ t_{x2} \circ t_{x3}(\mathbf{y}) \quad \text{and} \quad T_x^{-1}(\mathbf{z}) = t_{x3}^{-1} \circ t_{x2}^{-1} \circ t_{x1}^{-1}(\mathbf{z}) \qquad \text{eq. 2-26}$$

Note the reversal of the order of the transformations comprising the inverse, T_x^{-1}. Composition of transformations occurs in multilayer circuits and the reversal of the order of application in the inverse occurs physically as well as mathematically.

By habit we are used to reading the notation

$$T_x(\mathbf{y}) = \mathbf{z}$$
eq. 2-27

as meaning T_x is a given or fixed function. However in map-seeking circuits most often T_x is the entity being solved for. This will be denoted

$$\underset{\sim}{T_x}(\mathbf{y}) = \mathbf{z} \qquad\qquad \text{eq. 2-28}$$

In the simple example above in which y and z were triangles, the circuit would discover that they are related by a 135° right rotation T_x in the forward path and a 135° left rotation T_x^{-1} in the backward path, despite the fact that T_x and T_x^{-1} are only one pair out of many rotational mappings implemented in the circuit.

Of course T_x and T_x^{-1} could be far more complex mappings than simple rotation, and when each is the composition of two, three or more mappings,

$$\underset{\sim}{T_x}(\mathbf{y}) = \underset{\sim}{t_{x1}} \circ \underset{\sim}{t_{x2}} \circ \underset{\sim}{t_{x3}}(\mathbf{y}) \qquad\qquad \text{eq. 2-29}$$

the circuit's ability to discover the mapping relationship between two data patterns becomes very powerful, as will be seen throughout this book.

The three operating modes can now be simply defined:

Map-seeking mode is

$$\underset{\sim}{T_x}(\mathbf{y}) = \mathbf{z} \qquad\qquad \text{eq. 2-30}$$

In the map-seeking mode a data pattern y is applied to the forward path input, and a data pattern $\mathbf{z}$ is applied to the backward path input, and the circuit discovers the mapping pair T_x and T_x^{-1}, provided those mappings are among those implemented in the circuit.

Driven-mapping mode is

$$T_x(\mathbf{y}) = \mathbf{z} \qquad\qquad \text{eq. 2-31}$$

In driven-mapping, the circuit is provided y as the forward input and is instructed to apply T_x in particular out of all the implemented mappings, to produce $\mathbf{z}$ as the forward output. At the same time it automatically applies T_x^{-1} to whatever is on the backward input, say $\mathbf{z}'$, and produces $y' = T_x^{-1}(\mathbf{z}')$ on the backward output. If T_x and T_x^{-1} are the composition of mappings in several layers, then a mapping in each layer is driven.

Restricted-mapping mode is a variant of map-seeking in which the set of mappings available to be sought from is externally specified to be a subset or restriction of the set implemented in the circuit.

$$\underbrace{T_x^R}(\mathbf{y}) = \mathbf{z} \qquad \text{where } T_x^R \subset T_x \qquad\qquad \text{eq. 2-32}$$

The restriction may be "logical", in other words a pure set operation, or "fuzzy", in other words a probability distribution. In the latter case the prior probability of a mapping t_i is adjusted by the initialization of g_i.

Chapter 3
Shape Interpretation by Recharacterization

3.1 Shape Interpretation Mechanics

When alpine ski racers have to compete in heavily overcast conditions, particularly above treeline, the resulting flat light can make it completely impossible to discern the shape of the course surface, particularly when the racer is moving at high speed. In these conditions pine needles are often scattered on the course, and from these sparse dark markings alone the surface contours become easily visible to a moving racer. The absence of markings for a stretch can be very disconcerting. The ski race course scattered with pine needles is an unusually pure real world case of a surface interpretation solely by displacement of viewpoint. Ski racers' ability to perform at high levels in these conditions indicates how effective the human visual system is when called on to construct 3D terrain models using its shape-from-motion mechanism unsupported by other cues.

To recreate this pure terrain-shape-from-motion problem in a manner suitable for input to a map-seeking circuit, test scenarios are constructed consisting of a set of spheres of different diameters whose centers lie on an invisible virtual surface, as seen in Fig. 3-1 and Fig. 3-3.

When the illustrative surface is removed from the scene the spacing and size variation of the spheres provide no clue to the surface, as can be easily seen in the individual views, Fig. 3-1(b,c). In fact, the size clue prevails and induces a sense of depth which from Fig. 3-1(a) can be seen to be illusory. Though it cannot be demonstrated on paper, when the "ball terrain" is viewed initially from viewpoint 1 and the viewpoint is swung quickly through an angle even less than that between viewpoint 1 and viewpoint 2, the planar arrangement of the balls becomes immediately visible. When the motion of the viewpoint stops the sense of the planar arrangement vanishes in a fraction of a second and the size cues take over again, despite the fact that the observer already knows the balls are arranged in a plane. The effect is very dramatic.

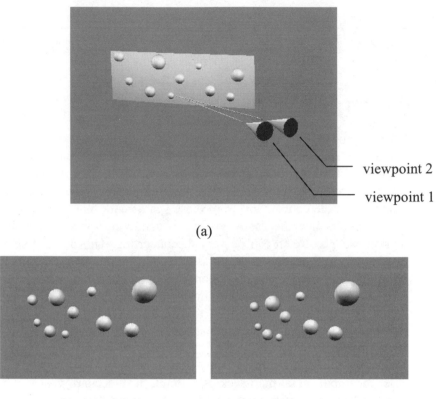

(a)

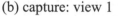 

(b) capture: view 1   (c) target: view 2

Fig. 3-1: Ball terrain example. (a) Geometry with illustrative plane; (b,c) actual views. $\Delta\psi$ = -8 degree (right rotation) viewpoint 1 to viewpoint 2. ψ = 40 degrees (angle between viewpoint 1 and plane).

A subtle but very important point can be deduced from the psychophysics of this demonstration. Whatever the nature of the internal 3D model being constructed the specific position of the balls is not part of it because it is impossible to recall the arrangement of the balls in such a way that their planarity is visible. Yet it is easy to recall the strong "sense" of planarity without recalling a specific arrangement of the balls. So the planar "sense" must be abstracted from the particular components that produced it. The conditions of the test make it certain that the planarity itself can only be deduced by the consistency of the perspective change across the set of balls as the viewpoint changes. There is no other information available from which to determine it.

A central point of this book is that shape is interpreted by using the identities of the transformations determined by map-seeking as a set of data for *recharacterizing* the input images. It will be seen that use of transformations as data, or *recharacterization*, turns out to be very powerful in explaining many aspects of human visual psychophysics. **The use of transformations as data rather than simply a means to establish "invariance" appears to be a fundamental and widespread operation in visual processing. It may be one of the core operations of the cortices, in other aspects of cognition and in motor control.**

Not only planarity but more complex shapes are made visible by viewpoint shifts. The undulations in the virtual surface of the ball terrain in Fig. 3-3 are as easy to see as the plane of Fig. 3-1 when the viewpoint is moved. Numerous traditional shape-from-motion experiments using moving dots also confirm this visual capacity to extract recallable shape without recalling the arrangement of the dots.[1, 2]

Initial Assumptions

The human visual system is capable of distinguishing local perspectivities and using them to determine shape. With multi-axis perspective mappings, changes in the transformation reveal inclination perpendicular to the line of view as well as parallel to it, but high resolutions are needed to discern transverse slope with accuracy. Even the low resolution map-seeking circuits used for demonstration in this book can provide rough data for calculating transverse slope, as will be seen later. But the first objective here is to demonstrate the simplest possible configuration that could serve the purpose. In the evolution of the visual system a simple, good-enough solution would have been both more likely and more robust than a complex, resource intensive, though mathematically complete solution. A limited set of low resolution single axis-perspective mappings would provide necessary and probably sufficient capability for a long legged creature running over uneven (but not cliff-like) terrain to judge the inclination of the surface ahead at various points. Such an assumption is consistent with the mono-tonic path of small adaptive steps with the highest probability of ascending to high levels of adaptive fitness.[3] In other words, the path to complex mappings requiring high resolutions goes through simple mappings requiring lower resolutions.

---

[1] Ullman 1979

[2] Prazdny 1980

[3] See Kaufman, Chapter 2 and 3 for the full discussion of this argument.

3.2 Geometry of Simple Terrain Angle Interpretation

In this chapter and the associated Appendices a number of trigonometric calculations will be presented which allow terrain geometries to be calculated from various mapping parameters available from map-seeking circuits. For simplicity most of these assume a narrow field of view from a relatively distant observer. Though these are presented in trigonometric form, this is not intended to suggest the brain contains trigonometric tables or function generators connected to compute the right-hand side of the equations. The purpose is to demonstrate that simple transformations applied to the mapping parameters yield the data necessary to build models of terrain from visual inputs. These transformations can themselves be implemented by map-seeking circuit, so the entire processing path from visual inputs to body-centered 3D model can be implemented by a sequence of map-seeking circuits. **The data which flow from one processing step to the next are transforms of the characteristics of the discovered mappings, not transforms of the image**.

The first of these calculations allows the slope of a terrain patch in the axis of the line of sight to be determined from the change in the mappings induced by a displacement and consequent rotation of the viewpoint. For simplicity the inclination transverse to the line of sight is assumed to be near zero, that is, near horizontal.

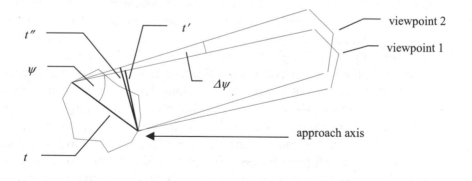

Fig. 3-2: Simple terrain angle geometry.

The line t is the projection of the approach axis on a patch of terrain view by a relatively distant observer from two viewpoints along the approach axis: viewpoints 1 and 2. The observer needs to determine the unknown angle ψ between the surface of the terrain and the line of view. The viewpoint shifts by a rotational angle $\Delta\psi$. This can be an oculomotor input, or determined from the translation of the pattern (on the retinotopic projection onto stage r) if there is no change in eye orientation, or from a combination of both. The line t' is the projection of t from viewpoint 1. The line t'' is the new projection of t from viewpoint 2. The shift in viewpoint causes a change in the length of the projection of t by a ratio *yscale*.

$$yscale = t'' / t'$$

The angle ψ is defined (arbitrarily) to be relative to viewpoint 1.

$$\psi = \tan^{-1}\left(\Delta\psi/\left(yscale - 1\right)\right) \qquad \text{(see Appendix D1)}$$

To determine *yscale* the circuit must have mappings with a range of foreshortenings in the direction of the axis of approach: for approximately horizontal terrain this is the vertical or y-axis. For a circuit with in-plane rotational mappings this could be either the y-axis of the input field or the y-axis of the memory field. The circuit used in the tests placed the *yscale* foreshortenings in layer 2, so that the y-axis rotates with the rotational mapping.[4]

Fig. 3-3(a) illustrates the setup for the "ball terrain" test. The sphere centers define an irregular surface viewed by an approaching observer from two positions. The circuit captures three fixations of view 1, as indicated by the view lines, and then finds each within view 2 to determine the *yscale*, and consequently the angle ψ for each fixation point.[5] Two of the view 1 fixations are laterally displaced on the far slope, roughly perpendicular to the

---

[4] In other words, if the head is tilted between the two viewpoints, the rotation mapping will accommodate it and align the y-axis foreshortening correctly in the second view. However, since only one axis of perspective is implemented, the memory view must be captured with its y-axis aligned as nearly as possible with the projection of the axis of approach. So if the head is inclined at time of capture, the compensating rotational mapping must be "driven" by vestibular inputs to rotate the input field into the correct orientation in memory. Perhaps this is one of the reasons for the optokinetic cervical reflex which attempts to keep the head level during banked turns. This is characteristic of humans, other mammals and birds. Pilots of high performance aircraft attempt to level the horizon even at bank angles beyond their ability to compensate (Merryman, Cacciopo 1997), suggesting that driven mappings as close as possible to vertical are preferred.

[5] If the viewer had been in motion and the fixations were sequential, each would have been from a slightly different view point. However, the speed with which humans determine terrain contours suggests there may very well be multiple mapping channels computing the *yscale*s and ψs for a number of points at each saccade. The case for this is even stronger in raptors, such as owls, whose eyes do not rotate in their sockets.

axis of approach. The "terrain" has a downward inclination to the left in this area, so the angle ψ_1 is greater than ψ_3.

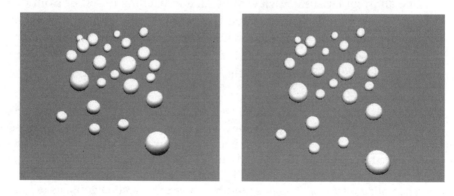

(a)

(b) view 1 (c) view 2

Fig. 3-3: Ball terrain test case. (a) Geometry with illustrative surface; (b, c) actual views. Three fixations from viewpoint 1: $\psi_1=29°$,, $\psi_3=25°$, $\psi_2=52°$, change in orientation viewpoint 1 to viewpoint 2: $\Delta\psi_1$, $\psi_3=5°$, $\Delta\psi_2=10°$. Surface shown for clarification in (a) not part of test data.

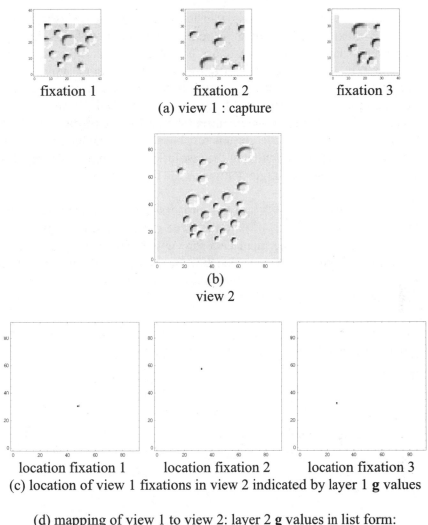

fixation 1   fixation 2   fixation 3
(a) view 1 : capture

(b)
view 2

location fixation 1   location fixation 2   location fixation 3
(c) location of view 1 fixations in view 2 indicated by layer 1 **g** values

(d) mapping of view 1 to view 2: layer 2 **g** values in list form:

rank	*map*	*theta*	*yscale*	*coeff*
fixation 1				
1	18	0.000	1.150	1.000
fixation 2				
1	17	0.000	1.100	1.000
2	18	0.000	1.150	0.024
fixation 3				
1	19	0.000	1.200	1.000

Fig. 3-4: Ball terrain test data. (Note: density plots inverted relative to image data.)

The circuit, as before, deploys translational mapping in layer 1 and rotational × y-axis foreshortening mappings in layer 2. That is to say, each mapping in layer 2 has a rotational parameter and a y-foreshortening parameter and so can be described by the pair < *theta, yscale*>. Those parameters are implicit in the identity of each layer 2 mapping, so the winning mapping can be translated back into its *theta* and *yscale* values. Theta increment is 10°, *yscale* increment is 0.05 and ranges from 0.5 to 1.5. No "zoom" scalings are needed in this test because the change in the viewpoint distance is too small to matter at the resolution of the internal representation. The change in viewing angle dominates the change in perspective.

The raw images, Fig. 3-3(b,c), are processed by edge filtering to produce circuit input data. Fig. 3-4(a) are the three initial fixations captured in view 1. Fig. 3-4(b) shows the input data for view 2. Fig. 3-4(c) are final layer 1 translation mappings showing location of the captured fixations in view 2. Fig. 3-4(d) are final layer 2 mappings in parametric < *theta, yscale*> list form for each of the three fixations.

The observed angles ψ_{1cir}, ψ_{2cir}, ψ_{3cir} are calculated from the oculomotor $\Delta\psi_i$ inputs and circuit generated $yscale_i$ mapping parameters.

$$\psi_{cir} = \tan^{-1}\left(\Delta\psi / \left(yscale - 1\right)\right)$$

$$\Delta\psi_1, \psi_3 = 5° = 0.0873, \quad \Delta\psi_2 = 10° = 0.1745$$

$$yscale_1 = 1.15$$

$$yscale_2 = 1.10$$

$$yscale_3 = 1.20$$

$$\psi_{1cir} = \tan^{-1}\left(0.0873 / \left(1.15 - 1\right)\right) = 30.2°$$

$$\psi_{2cir} = \tan^{-1}\left(0.1745 / \left(1.1 - 1\right)\right) = 60.2°$$

$$\psi_{3cir} = \tan^{-1}\left(0.0873 / \left(1.2 - 1\right)\right) = 23.6°$$

Graphically, this data can be summarized as shown in Fig. 3-5. The agreement between ψ_{cir} and the actual ψ is quite good.

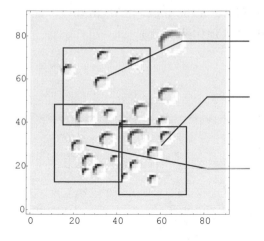

fix 2: $\psi_{2cir} = 60.2°$, $\psi_2 = 52°$

fix 1: $\psi_{1cir} = 30.2°$, $\psi_1 = 29°$

fix 3: $\psi_{3cir} = 23.6°$, $\psi_3 = 25°$

Fig. 3-5: Ball terrain ψ interpretation.

3.3 Effective Extent of Mappings

When the planar ball terrain of Fig. 3-1 is presented in rotation to a human viewer, the "sense" of the planarity extends to the entire virtual plane defined by the spheres. Does this occur because the viewer is fixating across all the targets (either concurrently by use of multiple channels, or sequentially by saccades) and discovering the angles calculated to be consistent with a planar interpretation? It is possible, but not necessary. The *effective extent* of a mapping, the area over which it correctly transforms the input image, can be determined by intersecting the input field with the backward mapped field. Where the mapping is correct the forward and backward signals will match, and the intersection of those two signals will be a good copy of the input pattern. Where the mapping is incorrect , the intersection will be by chance or not at all. This can be seen in the $r2_{fwd}$ intersect $r2_{bkwd}$ data, Fig. 2-9, in the cube test in Chapter 2. In the later frames the intersection has pruned the figure almost to the single face for that fixation and just a few ends of the other lines. This happens because the plane mapping that matches the two views across the extent of one face does not correctly map the other face. This illustrates that the extent of the mapping can be found by intersection, with only the scraps of chance intersections left around it.

Unlike the layer 2 intersection, the intersection of the layer 1 intersection $r1_{fwd} \cap r1_{bkwd}$ also locates the mapped extents in the input field, and is therefore more useful for interpretation. It is also the signal used for attention shifting. In Fig. 3-6 the layer 1 $r1_{fwd} \cap r1_{bkwd}$ data define the effective extent of each mapping in its proper place in the input field. In this example each represents the effective extent for a particular ψ_{cir}

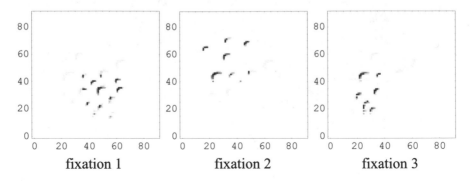

fixation 1   fixation 2   fixation 3

Fig. 3-6: Mapped extents in layer 1. Signal is $r1_{fwd} \cap r1_{bkwd}$.

3.4 Terrain Interpretation with Real Data

The ball terrain used in the previous demonstration establishes distinct "landmarks" which, though individually without identity, make the mapped matches unambiguous. Real terrain is not necessarily so cooperative. However, even homogeneous natural terrain offers variations which allow mapping to establish correspondence. After contrast enhancement, low pass filtering, thresholding and edge filtering, two displaced views of the same grassy slope, Fig. 3-7, provide many patterns for the matching to work on. The stability of the shapes of the intensity contours between the two views is sufficient to obtain useful mapping parameters even from small samples (40 pixel diameter, as usual).

(a) view 1   (b) view 2

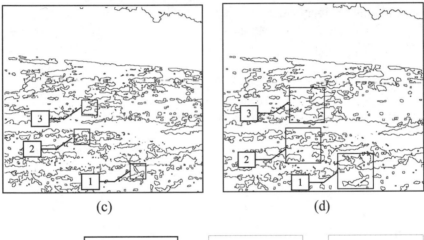

(c)   (d)

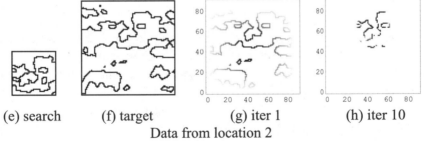

(e) search   (f) target   (g) iter 1   (h) iter 10

Data from location 2

Fig. 3-7: Slope angle determination from photo data.[6] (a) initial view; (b) view after movement; (c, d) initial and final input images after contrast enhancement, low pass filtering, thresholding and edge filtering; (e) location 1 captured sample 40x40; (f) location 1 target sample 90x90; (g) layer 1 intersection signal on iteration 1; (h) layer 1 intersection signal on iteration 10 showing mapped extent. Extent corresponds to about 2° field of view.

---

[6] Photo data courtesy Carnegie Mellon University test image archive

The final mappings for three locations in the scene are shown in Table 3-1. As in the ball terrain demonstration *yscale* is used to calculate the angle between line of view and the terrain ψ_{cir} at each location.

rank	map	orient	yscale	xscale	coeff
mappings for location 1					
1	2487	-2.000	1.060	0.930	1.000
2	3016	-1.750	1.060	0.930	0.790
3	2491	-2.000	1.060	0.970	0.784
4	2492	-2.000	1.060	0.980	0.776
5	3021	-1.750	1.060	0.980	0.770
mappings for location 2					
1	4562	-1.000	1.040	0.980	1.000
2	4033	-1.250	1.040	0.980	0.804
3	4561	-1.000	1.040	0.970	0.785
4	4056	-1.250	1.050	0.980	0.781
5	4055	-1.250	1.050	0.970	0.765
mappings for location 3					
1	4021	-1.250	1.030	1.090	1.000
2	5597	-0.500	1.030	0.980	0.669
3	6126	-0.250	1.030	0.980	0.652
4	5607	-0.500	1.030	1.080	0.564
5	5608	-0.500	1.030	1.090	0.551

Table 3-1: Layer 2 mappings for three locations

No ground truth or camera data are available for the images used in this test so approximation must serve for calculation of terrain angles. The camera vertical axis field of view is taken to be about 25° and is assumed to be inclined about 10° upward from the horizontal. Vertical displacement of each of the locations between the two views gave approximate value for $\Delta\psi_i$ of 0.63°, 0.31° and 0.21° for locations 1, 2 and 3 respectively. These values yielded ψ_{icir} of 9.5°, 7.7° and 6.8° respectively. Assuming a camera inclination of 10°, and approximating viewing angles to each location using the vertical distances from center of the image, the inclination of the slope along the line of view at each of the three locations is calculated to be

location 1	$\psi_{1cir} = 15.2°$
location 2	$\psi_{2cir} = 11.7°$
location 3	$\psi_{3cir} = 9.1°$

While these are reasonable values, the systematic inconsistency is more likely due to some combination of loss of effective resolution as distance increases and lens radial distortion, as evident from the trees at the top of the image, rather than actual slope change. The contributions of distorting factors cannot be reasonably determined from the available data.

The compression indicated by the average *xscale* (despite the shorter viewing distance for the second view) for locations 1 and 2 is consistent with the slight transverse slope downward to the left in the foreground. The calculation of transverse slope from multi-axis perspective mappings is discussed in section 3.5 below. The resolution of the image and circuit are insufficient to make more than indicative transverse slope calculations. The *xscale* and *yscale* data from location 2 (where the mapped extent represents a ground area large compared to the height of the grass), interpreted according to the geometry presented in Appendix D2, indicate the transverse slope would be about 1°. This is roughly consistent with the indications of transverse contours in the vicinity.

The orientation change between the two views of each location reflects both change in perspectivity and an approximately one degree axial camera rotation in view 2 relative to view 1, measurable from the change in angle of the boundary at the hilltop. The decreasing average orientation angle between views from location 1 to location 3 reflect the inverse effect of distance on perspectivity change. The greater variation in measured orientations and *xscale* values for location 3 probably result from the more acute observation angle and greater distance magnifying the effects of quantization. This suggests that the task requires averaging of many samples in each neighborhood and/or higher resolution both in the input image and in the circuit.

It should be recalled that the mappings implemented here are not true perspectivities but serve as approximations sufficient for the image resolution and the narrow visual angle of the mapped fields. The field covered by the mapped extent is roughly 2°, about half the human foveal field of view. It is sampled at about one third average foveal resolution.

Test Parameters

In order to make use of the available image resolution the mappings increments in layer 2 are small and the range of mappings is consequently limited: rotations between -3.0° and +3.0° by increments of 0.25°, x and y scaling between 0.9 and 1.12 by increments of 0.01. These parameters produce 13,225 maps in layer 2. The computation to find the mapping for each location takes about 75,000 dot products in the map-seeking algorithm, versus 33,000,000 by conventional correlation.

3.5 Determining Transverse Inclination, θ

To be useful in constructing a 3D model the description of a surface location must include orientation in two axes and a location in space. Perspective changes induced by motion can be used to obtain both ψ and the transverse inclination, θ. However, there are certain limits to this method, which will be discussed, and the visual system has several other ways of using map-seeking to determine θ, which will be discussed in turn. There are a number of other approaches to this problem including taking cues from transverse edges and shading, but this discussion is restricted to purely geometric aspects of the input image which may be determined by map-seeking circuits.

Multi-axis Perspective Transformations

Motion induces changes in perspective off the axis of motion which vary according to the slope of the surface perpendicular to the line of view. Such changes are distinguishable monocularly. Psychophysical tests using animated ball terrain displays suggest humans are quite capable of this discrimination. The discrimination is completely consistent with map-seeking circuits.

Mappings with foreshortening coefficients in both the vertical (y) axis and horizontal (x) axis of the view plane provide the parameters to determine transverse slope, θ, of mapped extents lying to one side of the axis of motion. Motion induces changes in the dimensions of a mapped extent in each axis in proportion to the slope. The determination depends both on relative changes in the two axes and the extent along each axis. The dimensions of the mapped extent are expressed as retinal angles, ϕ_{xi} and ϕ_{yi}, for view i.

$$\theta = \tan^{-1}\left(\frac{\sin \Delta\phi_x}{\sin \Delta\phi_y} \cdot \frac{\sin \phi_{y1} \sin \phi_{y2}}{\sin \phi_{x1} \sin \phi_{x2}}\right) \qquad \text{(see Appendix D2)}$$

where

$$\Delta\phi_x = \phi_{x2} - \phi_{x1}, \quad \Delta\phi_y = \phi_{y2} - \phi_{y1}$$

Once θ is determined, it is most effective to use this parameter to select the rotation that places the y-axis perpendicular to the transverse axis. This maximizes the use of the available resolution in determining ψ. This appears to be true also of the human visual system. With the ball terrain displays the sense of planarity is strongest when the viewpoint is rotated in an arc perpendicular to the plane of the balls. With the arc inclined even 30° from the perpendicular the sense of planarity is noticeably weakened.[7]

Mapping between Observed Projection and True Shapes

A variation on the use of full perspective to determine both ψ and θ does not require high resolutions, and therefore can be demonstrated more meaningfully than was possible with the data discussed in section 3.4 above. There are circumstances in which the angular change between the captured view and the current view are large enough so that both angles can be determined with useful accuracy even with low resolution datapaths. One circumstance that occurs frequently is that the visual system has previously captured a *canonical* [8] or true shape view of some aspect of an object. When that object first appears in a scene the view of it is usually oblique, thus the angular change between the canonical view and the current view is usually large. This is the same problem as the shape from motion problem, except the viewpoint displacement has taken place over an arbitrarily long period of time. Since the angular relationship of the viewer to the canonical view is standardized the mapping to the input field directly yields the orientation angles of the target.

We have already seen a "real data" instance of mapping to canonical shapes in the aerial photo interpretation in Chapter 2. Fig. 3-8 presents a "controlled" case. The circle, Fig. 3-8(a), has been memorized at some earlier time, and now the circuit encounters the inclined ellipse in the input field, Fig. 3-8(b). The layer 2 mappings in the circuit used for the canonical figure tests are parameterized directly in rotational angles ψ and θ, making the assumption of $\psi=90°$ and $\theta=90°$ for the canonical view. Both angle increments are 10° (though increments as small as 2.5° have been tested successfully.). The final layer 2 mapping here reads out both angles directly, rather than requiring input of the change in observing angle.

---

[7] This may be another reason for the optokinetic cervical reflex, mentioned in an earlier footnote.

[8] The term *canonical view* will be used to refer to a true plane shape, though in doing so it is extending the meaning from a somewhat different context. Canonical views can refer to 2D views of non-planar surfaces, which although not "true shapes" are standard, repeatable views of such surfaces. The canonical view of a face is straight on, from along the normal out of the centerpoint between the eyes.

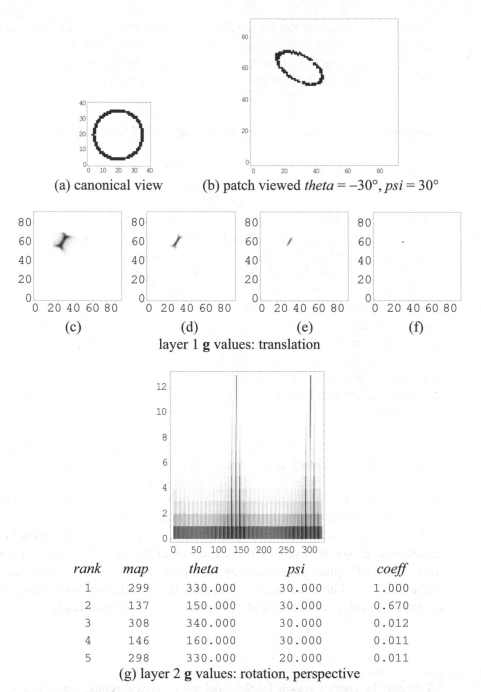

(a) canonical view　　　(b) patch viewed *theta* = −30°, *psi* = 30°

(c)　　　　(d)　　　　(e)　　　　(f)

layer 1 **g** values: translation

rank	map	theta	psi	coeff
1	299	330.000	30.000	1.000
2	137	150.000	30.000	0.670
3	308	340.000	30.000	0.012
4	146	160.000	30.000	0.011
5	298	330.000	20.000	0.011

(g) layer 2 **g** values: rotation, perspective

Fig. 3-8: Mapping perspective view to canonical view, ellipse.

Layer 1 **g** (translation) locates the canonical circle, now transformed into an ellipse, in the input field, and layer 2 **g** identifies the two axes of rotation that transform the circle into the ellipse. In this case the circuit has settled to two final layer 2 mappings. From the list presentation in Fig. 3-8(g), these two surviving mappings can be seen to differ by 180° in θ, but agree on 30° in ψ. It is the symmetry of the ellipse that produces the dual response: its face could be oriented downward as well as upward. Another temporary confusion caused by the symmetry can be seen in the layer 1 **g**. Until layer 2 resolves completely to only two mappings, layer 1 finds a number of mappings centered along the y-axis of the memory pattern of the ellipse.

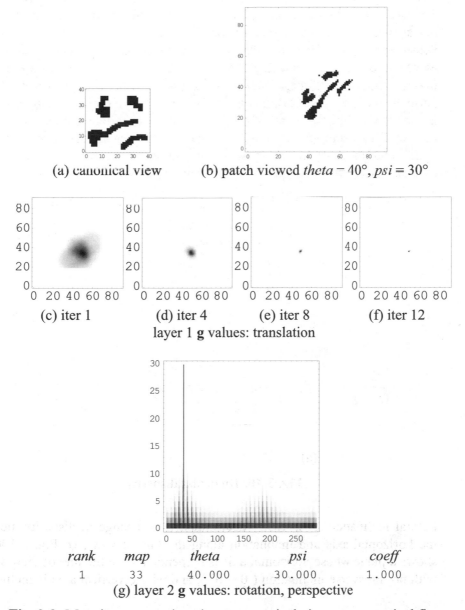

(a) canonical view   (b) patch viewed *theta* − 40°, *psi* = 30°

(c) iter 1   (d) iter 4   (e) iter 8   (f) iter 12

layer 1 **g** values: translation

rank	map	theta	psi	coeff
1	33	40.000	30.000	1.000

(g) layer 2 **g** values: rotation, perspective

Fig. 3-9: Mapping perspective view to canonical view, asymmetrical figure.

A similar viewing situation with an asymmetrical pattern produces only one final layer 2 mapping, as can be seen in Fig. 3-9(g). A short duration response at 180° from the winning mapping can be seen in Fig. 3-9(g).

Transverse Inclination from Binocular Disparity

For creatures with stereovision transverse inclination of a surface produces disparity differences along the horizontal axis of view (assuming the head is kept vertical). Humans are capable of fine discriminations and can decode at least two disparities in a single fixation if they start within the fusion limit of approximately 20′-40′.[9] Map-seeking circuits can directly measure disparity as the distance of horizontal translation between corresponding points in the views of the two eyes. Using this mechanism stereovision is a limited case of shape-from-motion, but where the two images are captured concurrently and the only mapping set deployed is horizontal translation (or strictly, horizontal and a very small vertical range of translation).

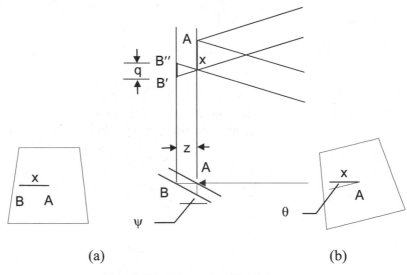

(a)   (b)

Fig. 3-10: Binocular disparity.

Lateral inclination, θ, is easily determined from change in disparity along the horizontal axis and inclination along the line of view, ψ. Fig. 3-10(a) shows a plane whose horizontal axis is perpendicular to the line of view and with no transverse inclination (θ = 0°), and whose vertical axis is inclined

---

[9] Marr, p 144

70

to the viewer ($0° < \psi < 90°$). The line of view is assumed to be perpendicular to the page. When the eyes converge on point A the disparity at point B is zero since the distance from the viewer to points A and B are equal. Fig. 3-10(b) shows the same plane rotated by $\theta > 0°$ (plane tips down to the left). The viewing distance to B has increased by distance z while the distance to A remains the same because A is on the axis of rotation. If the eyes remain converged on point A, disparity is produced at point B because it is projected onto the two retinas at different distances, B′ and B″, from point A.

The size of the disparity, q, is a function of ψ, θ, the interocular distance, e, the viewing distance, d, and the lateral displacement, x, between points A and B. For $d >> x$ and $e << d$

$$q = \frac{x \tan \theta}{\tan \psi} \cdot \frac{e}{d}$$ (See Appendix D3)

Applying the data for the ball terrain, and assuming $\frac{e}{d} = \frac{1}{20}$ (e.g. an animal with an interocular distance of 6 inches and a viewing distance of 10 feet), a displacement x of 20 units, and the minimum discriminable disparity q of 1 unit, then $\theta = \psi$. This means that for $\psi = 30°$ the minimum discriminable θ is also 30°, which is quite useless. Obviously the measurement of θ using disparity requires far finer resolution than the measurement of ψ (unless the creature has the interocular distance of a battleship rangefinder). The resolution of the stereopsis circuit must be about 5-10 times that of the shape-from-motion circuit to achieve similar accuracy in θ as in ψ, depending on typical viewing angles and distances.

It is interesting to estimate what various real creatures could discriminate in this case. With a viewing distance of about 150 units in the ball terrain example, a disparity $q = 1$ represents 23′ of arc. Marr estimates a resolution of about 1′ for a 20′ disparity[10] in humans. This is about 20 times the resolution of the demonstration circuit. This gives a minimum discriminable θ of 1.7°, quite a useful accuracy.[11] For horses, with angular resolutions about

---

[10] Marr, p 144
[11] Although no experimental psychophysical data could be found to confirm this figure, a simple experiment can be performed by the reader. Stand in the center of an asphalt covered road and look down at the surface about 10 feet distant. Choose a location where there are no cracks or lines in the road, so the random pattern of embedded gravel provides the only features. With two eyes fixed on a single point the slope of the crown is easily distinguished, even close to the center of the road. But with one eye the slope is much more difficult, if not impossible, to distinguish.

0.6 times that of humans, it allows a minimum discriminable θ of about 3°, a still very useful accuracy. For eagles, with angular resolutions about 4 times that of humans, it permits similar θ discrimination at twice the distance, given an interocular distance of about half that of humans.

3.6 The Need for a Viewpoint Independent 3D Model

David Marr, in his classic work "Vision," proposed a representation intermediate between the viewplane and a full 3D model which he called the 2.5 D model. It is a viewplane image of a scene augmented by vectors to indicate the orientation of the surface. In effect the 2.5D model is equivalent to what the map-seeking circuit has accomplished in determining ψ and θ for each mapped extent. But in the context of the problem of terrain interpretation this is not sufficient. To see why, consider a mountain goat preparing to leap across a void to a narrow ledge. He has to judge whether than ledge is long enough to land on, and how its surface actually slopes.

Fig. 3-11: The mountain goat's problem.

However the viewplane image, Fig. 3-11, of that ledge is very oblique and therefore very much shorter than its true length. To determine before leaping whether the ledge is long enough and level enough to land on and decelerate

---

The only source of data for this discrimination of lateral inclination is binocular disparity in the random pattern of the gravel.

the goat must transform that oblique projection into a reasonable approximation of a true shape and orientation. In other words, the goat needs to construct a 3D model, Fig. 3-12, to allow him to calculate the kinetics of the leap.

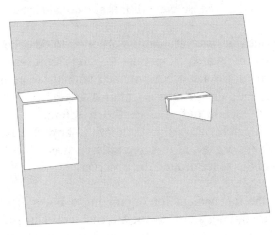

Fig. 3-12: The mountain goat's solution.

3.7 Floating Patch Representation

It is not known how any biological visual system stores 3D models of terrain or any other object whose surface it is interpreting. Psychophysical evidence becomes less constraining the further one moves from the external signal, so a wider range of hypotheses can survive the evidence. The best one can do is proceed using a set of reasonable assumptions to demonstrate that a path forward exists. Therefore the following sections addressing 3D representation are not part of the theory but rather presented to show that map-seeking circuits provide a means to construct *some* plausible 3D representation.

For the moment it may be useful to visualize a set of patches floating in space being assembled as the 2D visual system acquires the pieces of the scene and projects them into the proper location. This working assumption will be termed the *floating patch* representation. (A representation is not an encoding or a data structure, but an abstract organization for data which must be encoded and held in some physical medium to be useful.) We will assume a body-centered frame of reference: position and orientation relative to a coordinate system origin located in the torso. The penalty of this frame of reference is that location of the patches must be updated for self-motion, but its advantage is that it is the natural frame of reference for motor control.

Transformation of patches in the viewplane into floating patch representation involves two operations: locating the patch in the body-centered 3-space, and transformation of line-of-view relative orientation angles ψ and θ into body-centered orientations, which we will call ψ' and θ'. The second step depends on the first, so we will initially consider the location problem.

We can imagine the space around the observer to be encoded by a collection of cells, each of which responds to, and represents a limited three-dimensional field around it. Associated with these cells are other sets of cells which encode the surface orientation ψ' and θ'. A minimalist representation need encode only the center or reference point of the mapped extent for each patch. We will not commit to how these cells are organized, only that they are organized in some ordered way which permits translation and rotation of the space around the observer.

Fortunately, the mechanism for organizing a body of cells to represent a spatial area or volume, and to associate each cell with an ordered higher dimensional spatial input (e.g. oculomotor plus head position signals) has been provided by Kohonen.[12] A patch can be located in body-centered 3-space at point p by providing the high dimensional spatial inputs and allowing the cell representing p to respond. This cell (or cells if p is interpolated) is used as the input to next stage in two ways: to activate a location in the input field, and to enable the set of memory cells associated with location p to capture patch orientation.

The transformation of ψ and θ into ψ' and θ' is dependent only on the location of the patch relative to the observer, in other words the location p. The operation can be expressed

$$T_v(p) = b$$

where

$$v = <\psi, \theta>, \; b = <\psi', \theta'>$$

In this formulation each mapping T_v is parameterized by a view-centered orientation and it maps any location p into the correct body-centered orientation b. This can be performed by a circuit in the driven-mapping mode. In this instance it could be a single-layer circuit with p as the input to r_{fwd}, b encoded on b_{fwd} and b_{bkwd}, and a number of mappings each corresponding to one value of v. The b_{fwd} and b_{bkwd} datapaths are connected to the floating patch memory, but only the cell or cells which hold the body-centered ori-

---

[12] Kohonen 1988, Chapter 5

entation associated with the location *p* are enabled to capture the transformed orientation data.

The map-seeking mode of the circuit is engaged when the inverse transformation is needed. One task that requires the inverse mapping is determining viewer location from a new view and an already captured model.

It is not difficult to see how the circuit assembles an approximate model of the surface. If stereo depth measurements are available and accurate it can project patches into the model using viewing distance to each patch and convert the angles to body centered as described above. The resulting floating patch representation might look like Fig. 3-13.

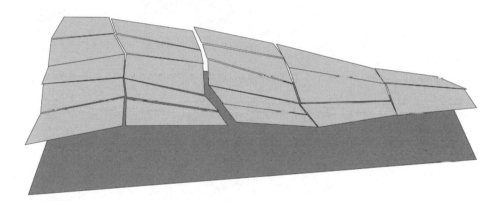

Fig. 3-13: Dense floating patch terrain representation.

Propagating ψ to Estimate θ

It may well be that a predator in pursuit, with eyes fixed on fleeing prey, must interpret terrain using image data from peripheral retinal areas with far lower than foveal resolution. This would preclude either accurate distance calculation or determination of θ by perspectivity or disparity unless the angles were quite large. In any event, past a certain distance, depending on the creature's visual acuity, stereo distance calculation becomes too inaccurate to use to project patches into place accurately enough to achieve a reasonably faithful model.

A smoother model may be assembled by starting at a point whose distance and orientation has been determined and then integrating outward patch by patch. As we have seen, ψ can be determined with low resolution pathways. Fortunately ψ and the direction of view are all that are needed to assemble an approximate model by integration. In the process, approximations of θ can be determined if they are not already available.

The ball terrain example, Fig. 3-14, provides a simple example of propagating surface calculations from one fixation to others. The three fixation points correspond to a triangle in three space, ABC viewed from point K. The vectors which connect vertices A to C and A to B can be computed from vectors which lie on the planes defined by the lines of view, KAC and KAB. In the viewer-centered representation assembled by the circuits, as we have seen, the projection of these vectors lie on the mapping extents of each patch. If A′B′C′ is the projection of ABC in the view plane, then A′C′ = **p′**+**q′** and A′B′ = **r′**+**s′**. Since the patches overlap, the vectors meet in the middle of the overlap.

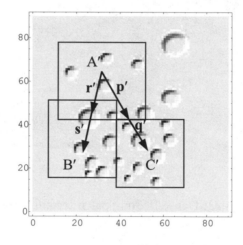

Fig. 3-14: Propagating ψ by mapped extents.

The projections of **p′**, **q′**, **r′**, and **s′** in the body-centered coordinate system are **p**, **q**, **r**, and **s**. Fig. 3-15 shows these seen from the side, in the plane formed by **p** and **r**. If A′B′ and A′C′ lie generally in the direction of motion, then ψ_B is a good approximation of the angle between **p** and **p′** and **q** and **q′**. Therefore the computed angles ψ_A, ψ_B, ψ_C can be used to project the lengths of **p**, **q**, **r**, and **s** from the lengths of **p′**, **q′**, **r′**, and **s′**. For the moment it is assumed the terrain is continuous so the ends of each pair of vectors are joined. The case of discontinuous terrain will be visited later.

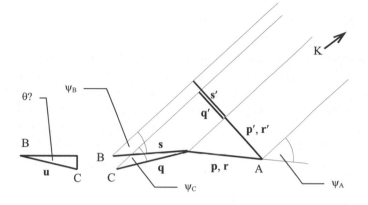

Fig. 3-15: Geometry of propagating ψ.

Expressed mathematically the rest is simple.

$$AC = \mathbf{p} + \mathbf{q} \,, \;\; AB = \mathbf{r} + \mathbf{s}, \;\; BC = \mathbf{u} = (\mathbf{p} + \mathbf{q}) - (\mathbf{r} + \mathbf{s})$$

The vector **u** can now be used to impart a transverse inclination to the two patches it touches. The θs for the patches may not equal the inclination of **u** in the transverse axis, but may need to be derived by interpolation or relaxation across more than two fixations, as will be seen shortly.

As constructed the patch containing the A fixation was considered to lie approximately perpendicular to the planes KAC and KAB, so **u** is computed relative to that. If the plane containing A is known to have a transverse inclination, that will be propagated to **u** through **p** and **r**.

The need to determine θ for the initial reference patch suggests that it will generally be located at or near a cue which can provide this. A transverse edge is a good example.

Implementation by Map-Seeking Circuit

What is the relevance of the vector description above to map-seeking circuits? So far the circuits have not been shown to perform such calculations. Consider the endpoint of each vector **p**, **q**, **r**, and **s** to be a particular mapping t_p, t_q, t_r, and t_s of its originating point: the mapping dependent on the vector's length and orientation. Then, in Fig. 3-15, the points A and C are related by

$$t_p \circ \quad t_q(A) = T_{p,q}(A) = C$$

This formulation will be recognized as the driven-mapping mode of the circuit and provides a simple mechanism for locating C given A, t_p and t_q. Reduction of the patch to a vector uses the fact that each patch is represented by a single point in space through which that vector passes. The vector length and orientation are provided by the mapped extent and orientation of the patch. So t_p and t_q are themselves driven by another circuit which maps from the true shape and orientation data associated with the patch location.

This sort of vector calculation is an important capability and it has application beyond the propagation of patch data. It is in fact a simple version of the forward kinematics of limbs. If one imagines each sequence of vectors (not necessarily in a single plane, as assumed above for simplicity) to be the segments of a limb, then the determination of sequential vector endpoints from the angles and lengths of the vectors is the same operation as the determination of joint locations in space from segment lengths and joint angles. As will be demonstrated in the next chapter, this calculation is performed very efficiently by map-seeking circuits in their driven-mapping mode.

If the representation of a surface by a sequence of radiating vector segments seems crude, consider that the finest shape-determining instrument we have is a sequence of limb segments radiating from a common root: the hand. The very first curved surface we learn by laying on of hands is the breast, and it may well be that we first learn to ground visually acquired curved surfaces in the encoding of metatarsal joint angles. In any event, we never lose the ability to make that transformation. If you view a distant mountain, it is quite easy to "feel" its contours under your hand.

Resolving Transverse Inclination

From the example of the hand, it might be concluded that transverse inclinations are not even needed to characterize a surface. However, in terrain interpretation it would seem essential for an animal to know the full inclination information of the surface he is about to land on. If only ψ is known for a patch then θ has to be initially treated as indeterminate. In a circuit implementation, indeterminacy is represented by initially allowing a range of plausible values for the parameter and allowing the map-seeking process to restrict this range, if possible. In this case a set of plausible θs are assumed and the resulting set of T_v are activated. The result is that a set of $<\psi', \theta'>$ values are projected for each patch. Fig. 3-16 shows a simplified cross-section of the terrain model showing multiple candidate orientations for each patch.

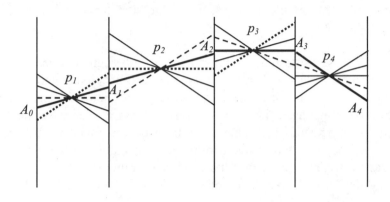

Fig. 3-16: Solving for transverse inclination.

The objective now is to prune each set of orientations constrained by the adjacent patches so that a consistent surface is left. As in the propagation of the patches away from the observer, each transverse vector is converted to a mapping of the initial point, A_i, to the endpoint, A_{i+1} , which lie along the planes denoted by the vertical lines in Fig. 3-16. The set of mappings t_i for each patch are constrained to lie on (or near) the point p_i so we will use the notation $t_i[p_i]$ to mean that the choice of mapping is constrained. Then in Fig. 3-16

$$t_i \left[p_i \right] \left(A_{i-1} \right) = A_i$$

The givens here are the p_i constraining the choice of t_i , as well as some prior knowledge of the range of locations of the A_i. The objective is to find the best set of t_i which create the smoothest surface.

$$A_4 = \underbrace{T_{4,3,2,1}}\left(A_0 \right) = \underbrace{t_4}\left[p_4 \right] \circ \underbrace{t_3}\left[p_3 \right] \circ \underbrace{t_2}\left[p_2 \right] \circ \underbrace{t_1}\left[p_1 \right]\left(A_0 \right) = A_4$$

This formulation is the map-seeking mode of a circuit with four layers, but the constraint conditions correspond to the restricted-mapping mode of the circuit, described earlier. The A_i begin as ranges of locations rather than single points, but these will be pruned as the competition proceeds. To enable the circuit to select the smoothest surface, the mismatch at each A_i needs to be minimized, but not at the expense of a very bad mismatch somewhere else. To achieve this optimization the activation, or strength, of the A_{i+1} field of endpoints decreases with increasing radius from the actual endpoint for the particular $t_i[p_i](A_{i-1})$. This means that the central mappings for each patch candidate are mapped most strongly by the next layer's mappings, thus favoring solutions that incorporate them. The circuit thus finds the path that has the strongest aggregate signal mapped through all of its layers.

Fig. 3-16 suggests there is kinship with limb segments here as well. The difference is that the segment lengths are not fixed and the joints are "fuzzy." But since the objective is to start with points in space and discover the mappings that connect them, this problem is otherwise identical to the inverse kinematics problem that will be demonstrated in the next chapter. As will be seen, the solution to inverse kinematics in the presence of constraints involves restricting the candidate points in each layer in a manner very similar to the restriction of the A_i as described above.

Practical Complications

The foregoing sketch of the 3D model assembly process, while it indicates that map-seeking circuits provide an end-to-end implementation, ignores a number of practicalities. A few of these are

1. In practice distant patches, generally viewed at more acute angles, span greater areas of actual terrain and so are not as accurate as those nearby. So while the nearby patches may be accurate enough to allow accurate high-speed foot placement, the further ones could only serve for route planning. A constant update must take place as the observer moves, to keep the nearby regions of the model sufficiently accurate. The floating patch representation permits such updating, but it requires some subtlety in establishing the proper relationships between newly captured patches and patches that have been previously part of the model.
2. Placing projections of patches into the 3D model by integration requires approximately contiguous boundaries. If boundaries are not contiguous the distance between patch centers must be estimated from other cues such as texture. The only way to accurately place isolated patches is to use distance and angle from the viewer, as described earlier.

The initial 3D representation might in fact be sparse, and look more like Fig. 3-17, than Fig. 3-13.

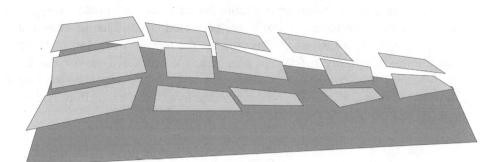

Fig. 3-17: Sparse floating patch representation.

The assumption of continuity is very strong in the visual system, as known from its natural behavior to extend whatever percept is not specifically terminated by a boundary or end feature. The question is where this takes place. Does it create continuity in assembling the 3D model? Or does it interpret a sparse 3D model to create the illusion of a continuous surface? The existing psychophysical evidence seems ambiguous. Can our mountain goat reliably complete his leap using a sparse 3D model? This puts the problem in very material terms: if the model is sparse, how does the goat interpolate between the isolated patches to determine a point in 3-space to use as input for the circuit performing the inverse kinematics for his legs? This is purely an engineering problem concerning the tradeoffs between accuracy and resources allocated to building and storing the 3D model. There are enormous economies in allowing the 3D model to be sparse.

Terrain Discontinuities

Discontinuities in terrain are also amenable to analysis from mapping parameters. A single monocular (or distant) view of a downward step away from the viewer may not contain any clue to its presence. For such *step discontinuities*, perspective differences resulting from viewpoint displacement provide sufficient information to characterize the discontinuity. We know from experience even a small movement of the viewpoint makes such a discontinuity very visible. It is easy to judge the height of such a step even without a view of the riser. The difference in translation of the terrain on the near and far sides of the discontinuity determines the height of the discontinuity relative to the viewing distance to the discontinuity.

Fig. 3-18 illustrates an example of a step discontinuity seen from the side. The objective is to calculate the height *h* of the discontinuity along the line of sight using the information available from the circuit. From viewpoint 1 two patches of the terrain, x and y, are projected as contiguous areas x' and y', and captured in memory. The view of the terrain to the left of the discontinuity is partially masked by the step. When the observer moves to viewpoint 2 the two patches of terrain are separately mapped by the circuit into two extents, x'' and y'', now separated by a'', the projection of the revealed section a of the terrain to the left of the step.

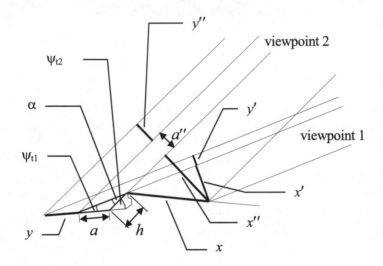

Fig. 3-18: Simple geometry of terrain discontinuities.

The angles of the surface of the left patch to the observer, ψ_{t1} and ψ_{t2} , are derived from the perspective determined by the circuit, as described earlier. The height of the discontinuity, h, can be computed as a proportion of the displacement a'' of the projection of the original visible patch of terrain, y'' from the projection of the patch x'' on the other side of the discontinuity.

$$\alpha = \psi_{t2} - \psi_{t1}$$
$$a = a'' / \sin \psi_{t2}$$
$$h / \sin \psi_{t1} = a / \sin \alpha$$
$$h = \frac{a \cdot \sin \psi_{t1}}{\sin \alpha} = \frac{a'' \cdot \sin \psi_{t1}}{\sin (\psi_{t2} - \psi_{t1}) \cdot \sin \psi_{t2}}$$

The shift, a'', is directly available from the layer 1 translation mappings. Each mapped extent in the view plane (e.g. Fig. 3-6) has an associated g_i which locates it in the input field. If two such views, displaced along the line of motion, are compared the g_i mappings separated by discontinuities will shift as shown in the diagram above. For small displacements of view the shift, a'', is the change in distance between the g_i on either side of the discontinuity.[13] Note that this takes three viewpoints to accomplish: one to capture the initial view, and two to obtain differently separated mappings for the two terrain patches.

Since h is proportional to a'', and on the retina the projected size of a'' is inversely proportional to the distance of viewing, the actual height of the discontinuity is computed as a proportion of the distance of the viewer from the discontinuity. This yields a more accurate measure of h than a direct comparison of the distances of the two patches from the viewer. For example, if a one-foot high discontinuity were observed from ten feet with a ninety percent accuracy in the distance estimate and the angle calculations, the computation from the displacement could be an inch or two off in either direction. By comparison, the direct measurement of distances could put it off by as much as two feet in either direction: a range of more than four times its height.

3.8 Floating Patch Representation of Objects

One critical virtue of floating patch representation is that it is isotropic. By projecting patches into true shape and then storing them with explicit angular rotations there is no loss of accuracy when the terrain steepens, or when the same system is used to capture the surface geometry of a face or a body or the hull of a schooner (Fig. 3-19). There is psychophysical evidence that humans at least store surface orientation in explicit angular terms.[14] And a simple test lends credence to that. The schooner of Fig. 3-19 can be readily oriented and rendered in a wide variety of natural and unnatural ways. Regardless of the manner of rendering the surface – wireframe, shading or texture – the discernment of the surface from small view shifts is unaffected by the orientation of the object.

---

[13] For large displacements of view the shift, a'', must be determined from the shift in the boundaries of mapping extents. This is because the change in perspective between two mappings with significantly different ψ alters the relative projected size of each.

[14] Stevens 1981

Fig. 3-19: Interpretation of 3D object surfaces. Wireframe view of model of schooner *Enchantress* (by the author).

Chapter 4
Unifying Vision and Kinematics

4.1 Grounding Vision

Consider the mountain goat's problem introduced in the last chapter. In deciding whether to leap there are obviously critical physical properties that are not computable from the image properties. The slickness of wet or mossy rock, poor footing on gravel, the uncertain solidity of a snow surface: all of these must be inferred by associating image properties with kinesthetic experience. But what about the geometric transformations we have been discussing? Are there aspects of these that also require grounding in non-visual experience?

If the needed sensorimotor transformational mappings have not been gifted to the goat in his DNA he must acquire them by experience. And this brings us back to the self-organization of the 3D model space by the process described by Kohonen. We had to step outside the purely visual to the motor to find a signal with which to organize the space. The training data for terrain interpretation mappings can be acquired by observing the same patches of terrain while moving about: something a goat can do in infancy. The calibration of the parameters of those mappings must use oculomotor and other motor inputs. And further, to calibrate the translation of those mapping parameters into surface angle, the latter must be determined by an independent measurement, such as the angular position of a hoof coming to rest on the surface.

4.2 Forward and Inverse Kinematics by Map-Seeking Circuits

The process of calibrating sensorimotor transformations between visual frames and body-centered frames of references is straightforward once the body-centered kinesthetic encoding of the spatial location of the hand, paw or hoof is known. The problem with a limb, however, is that its posture is first encoded as some function of muscle extensions specifying joint angular positions. Since these are in a many-to-one relationship with the information of interest – distance and angle from the frame origin (presumably in the trunk) to the end effector – some processing on the kinesthetic inputs needs to take place before establishing any association with the visual frame. This processing is known by the terms *forward* and *inverse kine-*

matics. Forward kinematics solves the problem of where in space a limb's endpoint (or end effector) is located given the angles of the joints and the lengths of the limb segments. Inverse kinematics solves the problem of what specific angles the joints must assume to place the limb's end effector at a given point in space. In traditional computation forward and inverse kinematics are solved by quite different procedures, the first simple, the second complicated. Constraints, particularly obstacles, add significantly to the complication of the second. By contrast, a single map-seeking circuit can solve both forward and backward kinematics in the presence of constraints and obstacles, as we shall see.

Simple Limb Kinematics

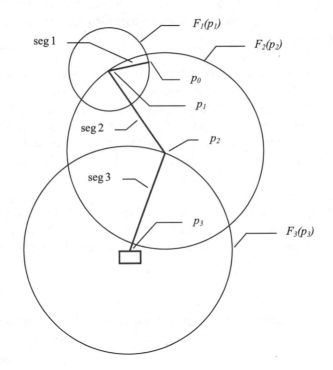

Fig. 4-1: Limb kinematics.

The set of possible positions of the endpoint (or distal end) of each segment of the limb is a mapping of the root (or proximal end) position of that segment. The mapping is a part of the surface of a sphere with radius equal to the length of the segment, centered on the root position. The constraints of motion around the root position define boundaries on the spherical surface. Let p_i be the position in space of the proximal end of the ith segment of the limb. Let $f_{i,j}(p_i)$ be the point on the spherical surface touched by the distal end of the ith segment when in the jth rotational orientation of the segment. In other words, $f_{i,j}(p_i)$ is the jth mapping of p_i.

Define $F_i = \bigcup_j f_{i,j}$

The points in space reachable by the ith segment from p_i is

$$S_i = \bigcup_j f_{i,j}(p_i) = F_i(p_i)$$

We will use the convention that the end effector is segment 1. For a three segment limb the points in space, S_1, reachable by the third segment from a given point p_3, at the root of the limb, is the composition of the mappings

$$\begin{aligned}S_1 &= F_1(S_2)\\ S_2 &= F_2(S_3)\\ S_3 &= F_3(p_3)\end{aligned}$$

or

$$S_1 = F_1 \circ F_2 \circ F_3(p_3)$$

A point in space p_0 is reachable from p_3 by the three segment limb if there exists a set of mappings $< f_{1,j1} \in F_2, f_{2,j2} \in F_2, f_{3,j3} \in F_3 >$ such that

$$p_0 = f_{1,j1} \circ f_{2,j2} \circ f_{3,j3}(p_3)$$

Applying this relationship to solve *forward kinematics* one supplies p_3 and the mappings $< f_{1,j1} \in F_2, f_{2,j2} \in F_2, f_{3,j3} \in F_3 >$, and from these one computes p_0. To solve *inverse kinematics* one supplies p_0 and p_3, and from these one computes one or more sets of mappings $< f_{1,j1} \in F_2, f_{2,j2} \in F_2, f_{3,j3} \in F_3 >$, or a null set to indicate that p_3 is not reachable from p_0. Using the notation to indicate solving for the mappings

$$p_0 = \underbrace{f_{1,j1}} \circ \underbrace{f_{2,j2}} \circ \underbrace{f_{3,j3}}(p_3)$$

These formulations of the problem are, once again, the *driven-map* mode and the *map-seeking* mode of a multilayer circuit. In this case, three layers starting from the input field deploy mapping sets F_1, F_2 and F_3 respectively. In other words, if the input field rl_{fwd} is provided the target position p_0 and the memory stage is provided the root position p_3, and the circuit is allowed to seek the mappings between them, the **g** of the three layers will converge on one or all of the mapping sets $< f_{1,j1} \in F_2, f_{2,j2} \in F_2, f_{3,j3} \in F_3 >$. To as-sure that only one set is determined a small amount of random noise is added to **g** to break any ties in the competition.

An additional complexity is imposed on this solution by the fact natural or man-made limbs do not have full rotational freedom. The rotational range of one segment relative to the one it is attached to is limited, and that limita-tion, in humans at least, varies with the angular position of the proximal segment relative to its proximal attachment. Restating this in the terms used above: the mappings allowed by one layer are constrained by the particular mapping selected by the layer representing the segment closer to the root of the limb. This would be trivially accomplished if each layer established a coordinate system relative to the current position of the segment below it. But this would be of little use since the end effector segment's coordinate system would then have to be transformed back into the coordinate system of the root, based on the particular mappings arrived at in the other layers. This could of course be accomplished by another circuit in the driven map mode, but a much more parsimonious solution is available.

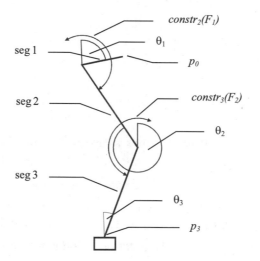

Fig. 4-2: Limb kinematics with constraints.

If all the mappings use the root's coordinate system, then each $f_{i,j}$ must pass a constraint to F_{i-1} based on the orientation j which in effect tells the next layer up that it can only choose from among a subset of F_{i-1} consistent with the rotational range of the joint between the ith segment and the i-1th seg-ment. Each active mapping in layer i sends a *constraint mask* to layer i-1 which is or-ed with all the other constraint masks, and the result filters the mappings considered in the candidate set for layer i-1. While this sounds very complicated and resource-intensive at first, that impression vanishes when one realizes that F_i is a small set: it has only as many members as there are rotational positions since each mapping $f_{i,j}(p_i)$ is defined for all p_i.

For example, each layer has 72 mappings in the two dimensional example to be demonstrated. In neural terms, as will be seen in Chapter 5, each mapping is implemented with two major cells. Projecting the constraint mask for each mapping to the next layer takes only one cell whose axon makes a synapse with each of the r-match cells (equivalent to q_i) in the next layer. In neural terms the constraint mechanism is extremely sparc.

Fig. 4-3 shows the operation of just such a circuit. The space used is a two dimensional version of body centered coordinate system assumed in the last chapter. Since we have created one grounded coordinate system, we might as well reuse it to tie the visual and kinesthetic systems together. (It might well be the other way round in biological systems: vision using the limbs' frame of reference.) The circuit has three layers, each of which deploys the mappings for one limb segment. The position of the target is applied to the input field, and the position of the limb root in the body's coordinate system is stored in a single memory. The circuit seeks a set of mappings which result in placing the end of the limb on the target. The rotational constraints of each joint are specified. This can be seen by the different solutions found when the constraints are altered, in Fig. 4-5 and Fig. 4-1.

The data in the panels is the activity on the r_{bkwd} path for each layer. The backward path represents the mappings from segment proximal to distal ends, while the forward path is the inverse mapping. Since the proximal to distal mapping is more intuitive, the data is more comprehensible. Each row shows the three mappings during one iteration of the convergence: from proximal (left) to distal (right). While both input and memory supply only a single active point, the intermediate layers accumulate the superpositions of all mappings of all projections of those points. The superpositions can be clearly seen in the data. The competition between mappings, however, prunes the mapping sets down to one per layer in 4-6 iterations.

Case1:
  Limb root and target positions

armx	armx	targx	targy
20	40	64	58

  Segment absolute angle limits, radius, and relative angle constraints

segment	θmin	θmax	$\theta incr$	radius	$\theta relmin$	$\theta relmax$
1:	−175.0	180.0	5.0	9	−175.0	0.0
2:	−175.0	180.0	5.0	23	0.0	180.0
3:	−80.0	80.0	5.0	31		

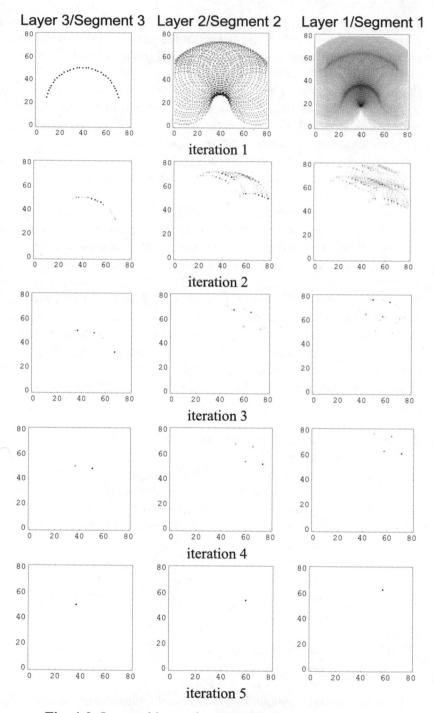

Fig. 4-3: Inverse kinematics mapping dynamics: **g** values.

The position of the limb is specified by the active **g** at the time of convergence, each specifying an angle defined relative to coordinate space (vertical is 0°). Final mappings in list form reading down from layer 1 to layer 3.

rank	map	theta	value
segment 1			
1	33	−15.000	1.000
2	34	−10.000	0.830
segment 2			
1	52	80.000	1.000
segment 3			
1	16	−5.000	1.000

Fig. 4-4: IK solution mappings, list form.

The solution position of the limb, plotted from the winning **g** parameters is seen in Fig. 4-4, below.

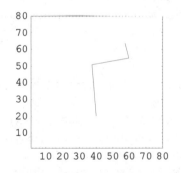

Fig. 4-5: IK solution, case 1.

When the relative joint constraints are changed, all else remaining equal, the circuit finds different solutions to the inverse kinematics problem.

Test parameters case 2:

armx	armx	targx	targy
20	40	64	58

segment	$\theta\,min$	$\theta\,max$	$\theta\,incr$	radius	$\theta\,relmin$	$\theta\,relmax$
1:	−175.0	180.0	5.0	9	−175.0	0.0
2:	−175.0	180.0	5.0	23	−175.0	0.0
3:	−80.0	80.0	5.0	31		

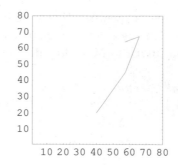

Fig. 4-6: IK solution, case 2.

In these demonstrations positions and mappings are discrete and the mappings are one-to-one. The solutions found above reflect this. To represent a three dimensional space with adequate resolution, it is far more efficient to use an interpolated encoding, wherein the relative activations of cells representing adjacent positions are interpolated to represent intermediate positions. Assuming just ten discriminable activation levels in each cell, a 3D space with one percent resolution can be represented by a thousand cells.[1] The mappings into such spaces obviously need to interpolate into these intermediate position encodings.

External Constraints or Obstacles in Kinematics

Joint rotation limits are not the only constraints imposed on a limb. Physical obstacles impose other constraints on the positions it can assume to reach a particular point. Fortunately, it is simple to implement obstacle constraints in a map-seeking circuit. The simplest solution is to remove from consideration any locations in the space occupied by obstacles. This is accomplished by forcing zeros in all occupied locations in each layer's r_{fwd} and r_{bkwd}. The circuit then seeks to solve the problem by mapping to locations not zeroed.

In the following tests the segments are allowed full relative rotation, and only the position and shape of the obstacle is varied. The root and target positions remain as in the preceding tests. No changes have been made to the circuits from the preceding tests. The location of the obstacle has been

---

[1] One percent is about 6.5mm in one arm span and is about the order of pointing repeatability without visual guidance, as the reader can easily confirm by blind pointing. Whether a one percent "grope" by our mountain goat is small enough to keep him from stumbling to his death is an interesting question.

forced to zero in all three layers, as can be seen in the r_{bkwd} data of Fig. 4-7 and Fig. 4-8. Only the first iteration is presented.

The test parameters for both cases:

armx	*armx*	*targx*	*targy*
20	40	64	58

segment	θ *min*	θ *max*	θ *incr*	*radius*	θ *relmin*	θ *relmax*
1:	-175.0	180.0	5.0	9	-175.0	180.0
2:	-175.0	180.0	5.0	23	-175.0	180.0
3:	-80.0	80.0	5.0	31		

Case 1: Obstacle at <x,y> =<45 to 55, 30 to 50>

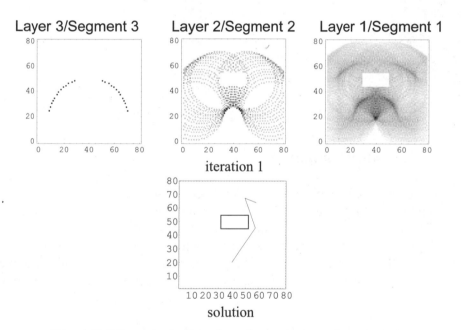

Fig. 4-7: IK test including obstacles and constraints, case 1.

Case 2: Obstacle at <x,y> =<35 to 56, 45 to 65>

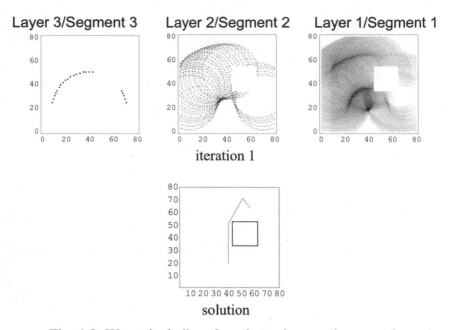

Fig. 4-8: IK test including obstacles and constraints, case 2.

Full Obstacle Avoidance

The approach to obstacle avoidance demonstrated above only forces the endpoints of the limb segments to locate themselves outside the obstacle. Though it did not occur in the test cases, it is certainly possible for the circuit to locate an endpoint in such a position that the segment itself would cross through the obstacle. This might be a useful and easily discovered evolutionary step, but it would lead to some bruises and tumbles at very least.

The map-seeking circuit solution for full obstacle avoidance uses a cooperating pair of three layer circuits: one for kinematics and the other for obstacle avoidance. In the avoidance circuit not just the endpoints, but the full extent of each limb segment (including limb thickness in three dimensions) is mapped into the space. The circuit operates in driven-map mode: the active **g** from the kinematics circuit drive the corresponding limb mapping **g** in the obstacle avoidance circuit. The r_{fwd} of each layer is forced to full value in any location where there is an obstacle. Since the backward maps are projecting the root position of each limb into all positions occupied by the limb, a collision between the limb and obstacle results in a non-zero

94

$r_{fwd} \bullet b_{bkwd}$ in calculating the **q** corresponding to that angular position of the limb. No competition is necessary, since that is taking place in the kinematics circuit to which the obstacle circuit is a slave. Any non-zero **q** in the obstacle circuit forces the corresponding **g** in the kinematics circuit to zero, thus removing it, and the orientation it represents, from the competition. The slave circuit is only put in operation when the kinematics circuit has pruned the possible **g** down to a few. If it is put into operation too early it could inadvertently eliminate a correct mapping that has been activated by a projection that has not yet been eliminated in the general pruning. Generally these are gone by the third iteration.

Biological Realization

A biological map-seeking circuit to implement limb kinematics, both forward and inverse, would reveal two populations of cells with distinct behaviors. The population corresponding to r_{fwd}/r_{bkwd} and b_{fwd}/b_{bkwd} encode positions in a body-centered space. The population corresponding to the **q** and **g** encode joint positions in several joint-local spaces. The cells corresponding layer 1 r_{fwd}/r_{bkwd}, encoding the target position, would show little or no sensitivity to limb posture. The cells corresponding to the layers intermediate between the root and the target would show activations in different postures except where the difference was an axial rotation and left the segment endpoint in the same location. On the other hand the cells encoding mapping coefficients, in other words encoding the joint position, would show different activations for all variations in pose, even those placing the segment endpoint on the same target location. At least one recent study using monkeys has identified cell populations in the premotor and motor cortices which are consistent with these categories.[2] Other experiments have localized sensorimotor transformations between visual and kinematic frames. This will be discussed further in Chapter 7.

4.3 Other Sources of Visual Information

So far we have seen map-seeking circuits applied to purely geometric transformations of images. There are other visual cues such as texture and shading whose initial discrimination is not geometric, yet can be transformed into geometric information compatible with the representations already established. In both cases an image characteristic is encoded as a local quantity or identity which varies in a systematic way with the orientation of the surface it defines. So the mapping between orientation and

---

[2] Kakei, Hoffman, Strick 2001

texture or shading involves changes in the encoding at each sample location. This adds a dimension to the transformations we have seen up to now: *feature transformations*.

We have encountered feature transformations in another form in the transformations between $<\psi, \theta>$ and $<\psi', \theta'>$. Because orientations are represented by distinct mappings, the transformation of one orientation into another is a change of identity rather than a change in quantity. In the last chapters the raw images have been processed by edge detection to form the input data for the circuits. In the human visual system it is likely that edges are encoded as specific visual plane orientations as a result of operations in the primary visual cortex area V1. It is easy to see that to rotate, or apply a perspective transformation to an image encoded this way, line segments of particular orientation must be mapped to a different orientation encoding when they are rotated spatially. This is a form of feature transformation, and it is just another transformational dimension that can be readily accommodated in a map-seeking circuit.

Shape from Texture

Though at a local scale texture is a geometric property, at that scale naturally occurring textures are generally not uniform enough to impart any useful information about an image. However, when observed at larger scales even natural textures have fairly stable spatial frequency characteristics. The spatial frequencies which characterize a surface are subject to the same geometric transformations as large scale features. If the visual system can make the assumption that the texture across a certain surface is reasonably uniform, the changes in the spatial frequencies of the texture map inversely to orientations relative to the line of view, just as perspective changes do, as we saw in Chapter 3.

This can be simply exploited by a map-seeking circuit to produce $<\psi, \theta>$ from texture gradients. The inputs to the circuit are a set of orientation specific, spatial frequency specific filters for each location. The set, or *feature vector*, is the *signature* for the texture at the location. For textures that lie on or close to the surface the transformation of the feature vector for orientation $<\psi_1, \theta_1>$ to the feature vector for orientation $<\psi_2, \theta_2>$ is the mapping between features whose spatial and orientation properties are related by the appropriate perspective transformation. That is, for each filter in the first signature there will be one in the second that reflects the spatial and/or orientation changes produced by the differences in surface angle relative to the line of view.

The extraction of $\langle\psi, \theta\rangle$ from texture therefore mimics the extraction of $\langle\psi, \theta\rangle$ from perspective due to motion or comparison with a canonical view, except here the mapping is from one point to another in the same view. The orientation of at least one point must be known from other characteristics. The signature of that point is used as the memory pattern. The discovered mappings between the reference point and all other points on the surface recharacterize the texture signatures directly into differences in orientation relative to the line of view for each location. This may take place concurrently, since only one memory is needed to serve as the reference. Ambiguities, which show up as multiple mappings in the same location, must be resolved either by propagating surface continuity constraints (as discussed in Chapter 3) from a known point, or using other visual information. As with perspective, observer motion induces a systematic change in the texture feature vectors. The use of these changes in building or augmenting a 3D model is similar to the use of perspective mapping changes.

Shape from Shading

In the ideal case of Lambertian illumination from a single light source, the changes in surface illumination are very similar in behavior to texture changes: there is a simple relationship of brightness to orientation that is equally amenable to mapping solution. In this case each feature vector encodes the illumination gradient and average value over a small area. This produces a signature that is subject to the same transformations as low relief texture. The signature of an area with known orientation serves as the reference and the orientations of other locations are read directly from the mappings between the reference and those locations.

For more complex reflectances, and multiple sources of illumination, the problem of shape from shading does not have a simple solution. There are two possible routes. The illumination on the surface as seen by the observer is the sum of a number of simpler illuminations, so the actual case may be decomposable into a superposition of a number of mappings, each corresponding to a simpler case. This has not yet been investigated. It is mathematically appealing, but it may not result in unambiguous interpretations. Alternatively, the simplest approach may be to simply learn mappings for various illumination conditions. Whether this approach is tractable has also not been investigated.

4.4 Terrain Vision Architecture

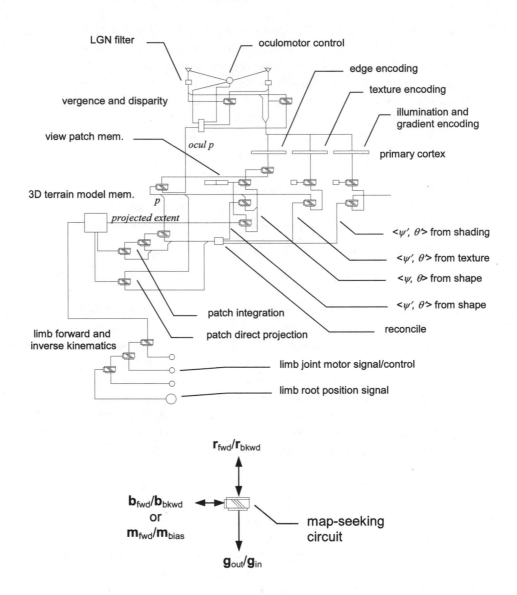

Fig. 4-9: Simple terrain vision block architecture.

If map-seeking circuits are the basic module of the visual system, how many are needed to execute a reasonably complete task, such as terrain interpretation? The flows discussed up to now can be roughly translated into a high-level architecture composed primarily of map-seeking circuits. A simplified block flow diagram, Fig. 4-9 summarizes the relationship of the various functions, and indicates the major data paths between map-seeking circuits which implement them. Fig. 4-9 is not presented for analysis but rather to convey an intuitive sense of the magnitude and complexity of a large scale visual system architecture built of map-seeking circuits.

Each of the nearly 20 rectangles marked with diagonal stripes represents a single or multilayer map-seeking circuit. The external r_{fwd}/r_{bkwd} paths emerge from the top of the rectangle, the external b_{fwd}/b_{bkwd} or m_{fwd}/m_{bias} paths emerge from the left side, and the mapping outputs **g**, and mapping initialization or bias paths emerge from the bottom. Control paths are not indicated.

Taking into consideration that advanced visual systems probably incorporate multiple channels (not shown) for most of these flows, the scale of a realistic visual system architecture begins to become apparent.

Fig. 4-9 also suggests the magnitude and complexity of the hardware needed to implement even specialized machine vision with biological scale capability. Using current technology it is possible to build a general purpose computational platform based on map-seeking circuits which would be well suited to investigating applications of this scale. This is discussed in Chapter 8.

Chapter 5

Neuronal Map-Seeking Circuits

5.1 Characteristics of A Neuronal Implementation

"All very well," intones the skeptical voice, "but so far I don't see anything here that looks like a brain." So far, true. For the map-seeking principle to have any possible relevance to neuroscience it must be shown to be realizable in circuits made of neurons organized as they are in the higher visual cortices. What we don't know about neurons and cortical circuits makes a conclusive demonstration impossible, but what we do know permits a convincing step in that direction. To guide the construction of a biologically plausible realization of the map-seeking circuit a list of axiomatic characteristics was distilled from the neuroscience literature.

The first seven observations are so common in the literature one dare not ignore them. Axioms 8 and 9 are less well established but convincing enough to be taken seriously.

1. reciprocal neural pathways (forward and backward) are essential in the computation
2. gamma frequency synchronized neural signal activity is an indicator of resolving or resolved computation
3. dendritic architecture (including spatial relationship of synapses) is significant in computation
4. single cell computation is integrating, non-linear and of limited range and precision
5. probability of interconnection between any two cells in a cortical area is low
6. learning takes place primarily in excitatory synapses
7. the computation must resolve in a relatively few cycles at gamma frequencies
8. increasing inhibition yields increasing order in the process of resolving the computation
9. cortical architecture for all higher sensory processing has a high degree of commonality

Relationship of the Neuronal Circuit to the Theory

The subject of this chapter is a circuit architecture composed of neuron-like elements which successfully implements the map-seeking behavior introduced and explored in the earlier chapters. The modeling of these elements is only intended to convey the minimal characteristics necessary to satisfy the axioms listed earlier, and in important aspects it is not realistic. The implications of the departures from reality will be discussed at length after the model is presented. The purpose of the model is to demonstrate that the map-seeking principle can be implemented by a circuit of elements with neuron-like characteristics, and to do this some very important concepts are introduced and tested which would be critical even in more realistic modeling.

At the level of individual neurons a wide range of modeling realism is possible. As yet it is not clear how much realism is essential to the function, and how much simply mimics the exigencies of the medium. Though it would be most convincing to model neurons at a detailed compartmental level, both computer resources and the author's time precluded this. At the other extreme, models of the simplicity of "neural net" neurons ignore axioms 2, 3, 4, 7 and 8. So the level adopted lies somewhere in between. All the axioms have been obeyed, but the neurons are unrealistic in a number of important respects which certainly might have been added to the list. A few of these are:

1. propagation time proportional to distance in dendrite and axon
2. stochastic synapse behavior
3. realistic variation of cell properties across the population

5.2 Cell Characteristics

In this implementation the cell is a generalized concept that permits a wide latitude dendritic structure and dynamic characteristics. The architecture of each class of cells is specific to its role in the circuit and differs significantly between classes. This is quite unlike the concept of a cell in "neural net." The synapses here have distinct functional characteristics depending on where they occur in the dendritic structure and where the cell occurs in the circuit. Synapses are excitatory or inhibitory. Only some excitatory synapses are capable of modification for learning. The threshold and saturation levels of different synapses vary according to function. Synapses are associated in groups called regions, and the regions may have distinct characteristics as well. The dendritic structures of cells may have one or more regions of synapses connected in a tree structure, the nodes of which may be additive or non-linear. The idealized implementation of the dendritic tree has no inherent delay. The cell body applies a sigmoid to the output of the dendritic structure and uses the result as the input to a leaky integrator. The sigmoid defines the threshold and saturation levels for the cell body computation. These are independent of the thresholds and saturation of the dendritic regions. The output signal rise and decay characteristics are governed by the leakage and gain. The axon here is simply a node with no structure of its own. The complete set of equations which implements this model appear in Appendix E.

There is no equivalent to an action potential or spike in this model. The rationalization for dispensing with spikes is that they are simply a biological expedient for efficient, low attenuation signal transmission, and they are converted by dendritic filtering into a smooth, longer duration pulse in which form they actually participate in most of the interactions with other signals impinging on the neuron. The simplification in this model was to leave the pulse in this form throughout its transmission.

In some examples of this chapter the behavior of an analog CMOS implementation of the circuit is used for demonstration because the higher resolution of the simulation generates more readable graphics. The device characteristics of the analog CMOS implementation yield a behavior similar to the idealized neuron. However, the axons and dendrites in the CMOS implementation have realistic capacitances which complicate their behavior somewhat.

5.3 The Cell Pair

A circuit module referred to as the *cell pair* forms the seed of all the circuits described. It provides generic forward and backward signal paths and the connections between them. A chain of n cell pairs serves as a basic module of repetition for a circuit of n stages, between which the connections can be particularized *en masse* by relatively simple specification. The reciprocal pathway within each cell pair defines the dynamics at the heart of the circuit's operation. Fig. 5-1 illustrates two stages of cell pairs.

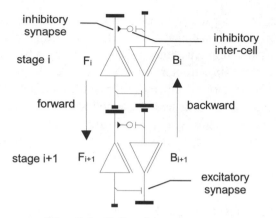

Fig. 5-1: Cell pairs, two stages.

The cell pair module provides

(a) an optional excitatory connection from the forward path cell to a proximal region of the backward path cell, and

(b) an optional inhibitory connection from the backward path cell, via a longer decay time-constant *inter-cell*, to a proximal region of the forward path cell. The inhibitory cell, when present, is referred to as part of the cell pair.

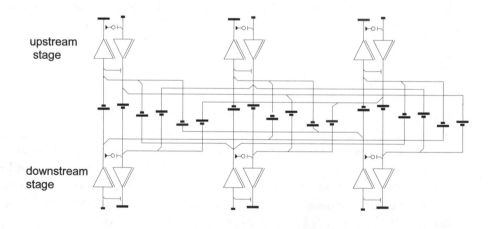

Fig. 5-2: Interconnections between two stages.

The cell pair as a basic unit is repeated to form a group or stage. The forward cell outputs or axons of a given group form synapses on the dendritic structure of one or many forward cells of zero or more downstream groups. The backward cell outputs or axons of a given group form synapses on the dendritic structure of one or many backward cells of zero or more upstream groups. (Upsteam and downstream are defined relative to the forward path.)

Fig. 5-2 illustrates two stages fully interconnected. This figure makes evident the problem of rendering connection between stages. From here on large multi-stage circuits are drawn in a format which orthogonalizes the inputs and outputs of successive stages and thus makes the interconnection patterns more readable.

Oscillatory Dynamics in a Chain of Cell Pairs

Circuits constructed of cell pair chains perform their computations primarily in the temporal domain. Differences in quantities are represented as differences in the arrival times and rise times of pulses or waves. The propagation of a pulse through a series of cells depends on the time constants of the cells and the timing and shape of the signals arriving on the synapses of each cell in the chain.

One way to exploit this is to configure a chain of cell pairs as an oscillator by setting its most downstream forward to backward excitatory connection and its most upstream backward to forward inhibitory connection sufficiently strong. If an input is maintained on the most upstream forward cell, distal to the inhibitory connection, a series of pulses will propagate around the loop. The periodicity of the pulses is governed by the strength of the driving signal, the gains and propagation time around the loop, and the decay time of the inhibitory signal (via the inter-cell) shunting the signal to F.

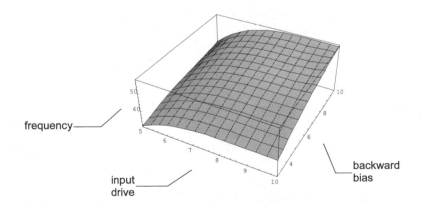

Fig. 5-3: Frequency versus drive.

Fig. 5-3 illustrates the observed frequency of a four stage cell pair chain for variations in input drive to the forward cell of the first stage and backward bias applied to the backward cell of the last stage. The backward bias plus the signal coupled from the forward to backward cell of the last pair comprise the backward drive of the chain. The monotonic relationship of frequency to the forward and backward drives is essential to the computational behavior of the circuits assembled from cell pair modules.

It must be kept in mind that the frequency is only a manifestation of the rise times of each cell along the loop. ***The circuit does not require its oscillatory behavior to arise from the looping connection in the first stage: it could as well arise from an external oscillator.*** The oscillator's frequency could be modulated by a signal from the backward path, but it isn't necessary. Even if initially gated by a fixed "clock," differences in input activation and internal circuit signals result in differences in the rise times along the path, resulting in differences in the arrival of the pulse leading edges which the circuits use to perform their computations.

5.4 Circuit Architecture

Using the cell pair as a module, the algorithmic circuit of Chapter 2 is readily translated into a circuit composed of neuronal elements.

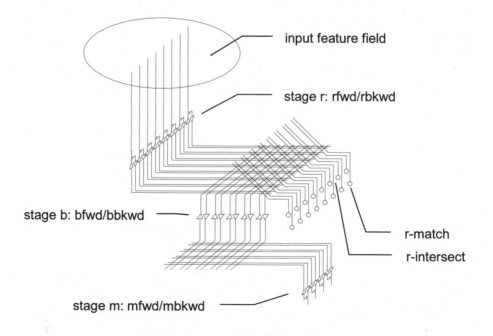

Fig. 5-4: Single layer neuronal circuit. (See footnote 1).

A single layer neuronal circuit, Fig. 5-4, is composed of three stages: r, b and m, (for *receiver*, *buffer*, and *memory*), each consisting of a population of cell pairs. One cell of each pair implements a stage in a forward pathway, and its associate implements a stage in a backward pathway. The pairing of the pathways by cell pair contrasts with the architecture depicted for the algorithmic circuit. This is more than a presentation issue. The physical construction of a cell-based circuit makes the proximity of the reciprocal pairs advantageous in stage r and m where the members of the pair are interconnected, and at the interconnections of stage r and stage b where the one signal, *r-intersect*, controls the forward and backward mappings and where the match between associated *rfwd* and *bbkwd* [1] signals are computed pairwise in the dendritic regions of *r-match*.

To give Fig. 5-4 an anatomical interpretation, the mesh of interconnections between stages would reside in the neuropil and the cells would occupy various layers (in the anatomical, not circuit architecture sense) below the cortical surface. Of course the forward and backward pyramidal cells are oriented in the same direction in the cortices, not reciprocally as illustrated here, and their axons return toward the cortical surface to make most of their connections.

Circuit Details

The details of a working neuronal circuit appear in Fig. 5-5, with a key to the symbols in Fig. 5-6. The backbone of the circuit is simply constructed by replication of cell pair chains, with their reciprocal pathways and distinct stages. Mapping circuitry and its control are added to this backbone.

Both triangles and circles designate cells. The dendritic arborization mentioned earlier does not appear in this diagram, but an important aspect of the dendritic structure is indicated by lengths of bold line connected by fine lines. These are particularly evident in stage m of Fig. 5-5 and in the circuit key in Fig. 5-6. The bold line sections are called dendritic *regions*, and designate localized character and interaction for the synapses impinging on the region. The localized processing within regions, and the often non-linear combining of outputs from regions, give rise to the powerful processing achieved in the dendritic structure in this theory.

---

[1] To distinguish between neuronal and algorithmic circuit pathways the subscripted notation of the algorithmic circuit is replaced by non-subscripted labeling for neuronal circuits.

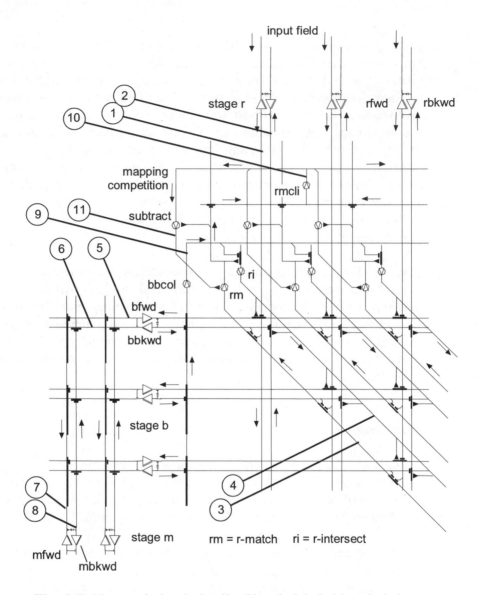

Fig. 5-5: Neuronal circuit details. Signals labeled by circled numbers are listed below, identified by algorithmic equivalents.

(1) $\vec{r}_1$ on *rfwd*   (2) $\bar{r}_1$ on *rbkwd*

(3) q_1 on *rm*   (4) g_1 on *ri*

(5) $\vec{b}_1$ on *bfwd*   (6) $\bar{b}_1$ on *bbkwd*

(7) $\vec{m}_1$ on *mfwd*   (8) $\bar{m}_1$ on *mbkwd*

(9) $\max \overleftarrow{\mathbf{b}}$ on *bbcol*   (10) $\max \mathbf{q}$ on *rmcli*

(11) $\max \mathbf{q} - q_1$ on *subtract*

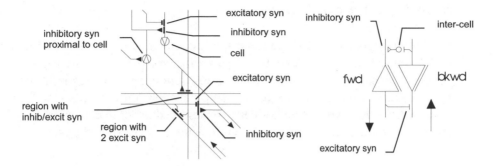

Fig. 5-6: Key to neuronal circuit diagrams.

The sections of fine line connecting the regions imply an arborization which cannot be rendered without impossibly complicating the diagram. In the formal specifications from which the simulated circuits are compiled, the actual arborization is specified in the dendritic combining function, *dcf,* as seen in Appendix E, equation (iv). Within the tree structure, nodes joining different groups of regions often have different processing characteristics.

In the discussion below both signals and pathways of the neuronal circuit are referred to by the same label, which evokes the equivalent path in the algorithmic circuit. The equivalence often helps in explaining the operation of the neuronal circuit. Thus *bfwd* refers to both the dendrite and axon of the stage b forward cells.

Also to simplify the discussion, some of the computations performed by the cells are referred by their algorithmic equivalents. It must be kept in mind that the cells do not perform exact equivalents of those numerical operations. For example, the matching in r-match cells is performed by first applying a non-linearity to the pair of signals impinging on each two input region, in effect roughly *and*-ing or "multiplying" them, and then summing the outputs of groups of two-input regions. These group sums are combined non-linearly in a manner similar to the "multiply" just described. Roughly, the cell is computing a product-of-sums-of-products, so the dot product representation of this operation is quite an oversimplification. The memory forward cells compute a dot product of the synaptic weights and synaptic inputs to each region, but then applies a threshold to each region before summing the output of all regions. Here too a simple dot product of the entire weight and input vectors is an over-simplification.

From input feature fields to stage r

The visual input to stage r (the input field) is assumed to be a representation of the retinal image processed into sets of characterizing features, as described in Chapter 2. The feature sets are *retinotopically* mapped onto stage r, meaning that the geometric relationship between retinal cells which ultimately drive the feature detector cells is preserved in some consistent way in stage r. Each feature provides an input to a single forward cell of stage r.

From stage r to stage b

The outputs or axons of stage r forward cells, *rfwd,* form a set of very comprehensive interconnections with the inputs or dendrites of stage b forward, *bfwd,* cells. Visual transformations are implemented by inhibiting or otherwise restricting the action of all but a few of the interconnections between stage r and stage b so that those few interconnections implement a map from a subfield of stage r forward (corresponding to $d(\vec{r})$) *onto* stage b forward, and an inverse map from stage b backward (corresponding to $d^{-1}(\vec{b})$) *into* the corresponding subfield of stage r backward (corresponding to path *rbkwd*). (*Onto* and *into* are used in the mathematical sense.)

Sets of interconnections between stages r and b implement each of these mappings and are controlled by the output or axon of a single cell belonging to a population of *r-intersect* cells (corresponding approximately to g_i), identified in Fig. 5-5 as *ri*. Associated with each r-intersect cell is an *r-match* cell (corresponding to q_i), identified in Fig. 5-5 as *rm*, which responds to goodness-of-match between the mapping of the stage r forward subfield and stage b backward. Each pair of *ri* and *rm* cells implements a mapping.

In the neuronal circuit presented here *ri* provides an input to the inhibitory synapse of a number of two input regions on the dendrites of the *bfwd* or *rbkwd* cells. The other input to each of these regions is excitatory and comes from the source gated by *ri*, either *rfwd* or *bbkwd*. If the inhibitory input from *ri* is present, the region's output is always zero. If the inhibitory input from *ri* is absent, the region's output is equal to the excitatory input to that region. All the regions are summed, thus forming a superposition of the source inputs *not* inhibited by signals from the associated *ri* cells.

From stage b forward to stage m to stage b backward

Memory, stage m, is a population of cell pairs, *mfwd* and *mbkwd*, each of which contains a single memory. The dendrite of each *mfwd* receives inputs from all the *bfwd* axons. To capture a signal pattern in memory, the synapses of one *mfwd* are adjusted to render a maximal response to the signal pattern being applied from *bfwd*. The synapses between the associated *mbkwd* and all the *bbkwd* cells are adjusted at the same time so a replica of the trained synaptic weight pattern, scaled by the output of *mbkwd*, is projected onto *bbkwd* when *mfwd* is active. The dendritic structure of *bbkwd* sums all the active inputs from the *mbkwd* cells, forming the *bbkwd* signal superposition.

In memory we encounter one of the divergences from the algorithmic circuit. In the algorithmic circuit the match between signal and a memory is computed as the dot product of $\bar{\mathbf{b}}$ and $\mathbf{w}_k$ (the synaptic weights) for that memory. In the neuronal circuit the match is computed in local sections, each corresponding to one or more dendritic regions, and the results are combined. The dendritic structure can be arranged so that it rewards good matches between inputs and weights which occur on sections of the dendrites which correspond to contiguous local areas in the input field. The region outputs are thresholded and the results of these local matches are combined in a tree-like structure whose nodes are functions which may be nonlinear. Consequently the dendritic structure punishes scattered matches because the outputs from local sections are lower than they are when the local matches are strong . This non-linear character of the stage m forward cells (and similarly for r-match cells), combined with the threshold in the sigmoid of the forward memory cell body – plays an important role in the robustness of the circuit.

An additional complication of this dendritic non-linearity allows the various regions of the dendritic structure to be combined with varying "strictness" ranging between *or* and *and* depending on an applied bias. This is presented in more detail in Appendix E.

And back to stage r

The reciprocal pathway from stage m backward, through stage b backward and stage r backward finally rejoins the forward pathway at stage r with an inhibitory synapse via the inter-cell. This inhibitory junction closes a loop, which results in oscillatory response to the input signal even if the input signal is DC, as discussed earlier in the section on the cell pair. DC inputs are used in all of the test cases below.

Competition

The assemblage of cells in Fig. 5-5 to the right of the label *mapping competition* implement a variant of the *comp( )* function presented in Chapter 2. Signals in the neuronal circuit encode amplitude by phase (*more* is represented by *earlier*), so an operation which outputs a pulse coincident with the first wavefront of a group implements the *max( )* function. The cells labeled *bbcol* and *rmcli* implement this function. Their thresholds, clipping levels, and time constants are chosen so the first input produces the output pulse, and the timing is relatively unaffected by multiple early arriving pulses.

The output of *bbcol* generates a signal used to activate r-intersect (labeled *ri*) cells and thus block the gating of *rfwd* signal onto *bfwd* and *bbkwd* signal onto *rbkwd*. Thus all such gating is blocked unless the *bbcol* signal itself is inhibited from activating a particular *ri* cell. That inhibitory signal comes from each associated r-match (labeled *rm*) cell. The outputs of the *rm* cells are themselves attenuated by a signal from the competitive circuitry formed by *rmcli* and the *subtract* cells. The *subtract* cells in effect subtract the associated *rm* output from the signal produced by *rmcli*. Later and smaller *rm* signals diminish the *rmcli* signal less, thus implementing $\max \mathbf{q} - q_i$. This, it will be recalled, is the heart of the *comp( )* function in Chapter 2.

Unlike the algorithmic circuit, in which **g** keeps the state of the competition, in this version of the neuronal circuit *rm* keeps the state. Therefore the results of the subtraction are used to inhibit the associated *rm* cell. The time constants of *rm* and *subtract* are chosen so that the reduction in output of *rm* is carried over from cycle to cycle.

The process of competition gradually reduces the output of all but the *rm* cell with the strongest inputs from the matching regions of its dendritic structure. Thus only the *ri* cell associated with that winning *rm* is inhibited, allowing gating of its mapped signals from *rfwd* onto *bfwd* and *bbkwd* onto *rbkwd*. All the other *ri* cells are active, thus blocking the gating of their mappings. (That is why the outputs of the *rm* cells, rather than *ri* cells, are used to illustrate the state of the mapping in neuronal circuits.)

The time constant of *ri* is set to distinguish between the *recognition* and *non-recognition* states: i.e. the decay of the *ri* signal is long compared to the recognition frequency of the circuit. Therefore, if at the end of competition none of the *ri* cells is inhibited by the output of an *rm*, all gating onto *bfwd* is inhibited for the duration of the *ri* decay. That is the non-recognition state.

The relationships of *rm*, *ri* and *bbcol* for recognition and non-recognition states can be seen in Fig. 5-7.

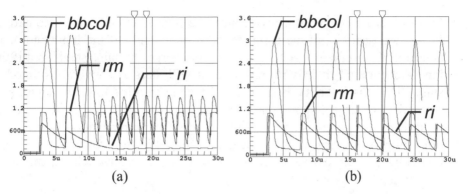

(a)   (b)

Fig. 5-7: Signal relationships. (a) recognition state: *rm*, *ri*, *bbcol*; (b) non-recognition state: *rm*, *ri*, *bbcol*. (CMOS model, time in usec).

Numerous variants of the competitive circuitry are possible. The *ri* signal can be made excitatory instead of inhibitory, and the "logic" of selection reversed. This allows all gating synapses on the *bfwd* and *rbkwd* regions to be excitatory, which may be more biologically realistic. (This variant has been tested and works equally well, but requires one more cell per mapping to gate the activating signal to *ri*.) Many variants on the time constants and threshold levels of the various cells are also possible.

Several mechanisms associated with learning are not shown in Fig. 5-5. An external input allows forced activation of a single *ri* so that a single input subfield is mapped to train memory. Another set of inputs selects which memory cell pair is trained by the signal pattern on *bfwd*. A small amount of circuitry synchronizes the training "enable" signal to the *bbkwd* cells to capture the signal from the associated *bfwd* cells when they are spread out in the quasi-linear regime prior to saturation.

Connectivity and Scale

Since each mapping touches only a small subfield of stage r and there are no interconnections within a stage the probability of interconnection between any two cells is quite low. For the circuits used in testing the probability of interconnection is about 0.2% for all cells, and about 1.4% for "datapath" cells. These interconnection probabilities are of the same magnitude as encountered between pyramidals in primate cortices.

5.5 Circuit Dynamics

In the algorithmic circuit, convergence to a final state ends the computation. In the neuronal circuit the computation doesn't end. If the inputs are kept constant, the oscillatory state to which it converges remains stable, but if the inputs change when it is in one state, it will start to converge to another state. This characteristic is responsible for the circuit's inherent nature as a recognizer of targets in motion. When targets move with respect to the input field there is a progression of activity across the *rm* population as the mapping follows the target, as will be demonstrated in the next chapter. Static recognition is the particular case in which the active *rm* remain the same over time.

The temporal aspects of the neuronal circuit start with its means of encoding quantities which is then exploited to implement all the computations it performs.

Temporal Encoding

It can be seen from the above discussion that if a number of mappings and memories are expected to be active at the beginning of the competitive process, the sums of the signals on stage b may be very large compared to the single signal of the final stable recognition state. Neurons and CMOS transistors have very limited quasi-linear regimes above a fairly high threshold. How then to accommodate the great dynamic range required to implement the circuit? The solution lies in a form of temporal encoding in which magnitude is represented by relative phase or delay of the rise of the leading edge of the signal. All addition or comparison of signals is accomplished during the risetime of the signal within the quasi-linear regime of the cell or device. When summing two signals of different phase, at any moment during the rise of the leading signal the lagging signal is contributing less (or nothing) to the sum. Fig. 5-8 illustrates an example of temporal encoding in a CMOS circuit in which the lag is non-linearly proportional to DC input voltage difference.

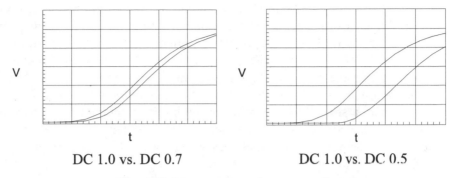

DC 1.0 vs. DC 0.7   DC 1.0 vs. DC 0.5

Fig. 5-8: Temporal or phase encoding.

When many signals are contributing to a sum the aggregate moves into the quasi-linear regime when all its contributors are very small. When just a few signals are contributing the aggregate enters the same quasi-linear regime when the contributors are larger. A signal with few contributors therefore rises later relative to its contributors than a signal with many contributors. Using the time domain to encode amplitude vastly broadens the effective dynamic range of the circuit much the way a floating-point exponent increases the effective number range of a given binary word length. Usually all the contributors to a given signal are operating in the same neighborhood of the range at a given time. But even when they are not, the laggards are by definition small, so even if their arrival is delayed until the strong early contributors have driven a device into clipping the resulting inaccuracy in the computation is not significant. The analogy to floating point normalization is evident. The leading edge of the front of signals on a group of pathways starts the computing of any process that affects all of them. This self-normalizing effect accomplishes the same result as the scaling step in the algorithmic circuit.

The nature of this temporal encoding does introduce a monitoring difficulty in that the waveforms of signals representing widely different magnitudes may appear identical. Often their difference appears only in their relative phase until inhibition is applied to suppress the laggards. Therefore a signal can only be monitored meaningfully in relation to its peers and in relation to any inhibitory signal used to suppress it.

Advantages of Oscillatory Dynamics

Even with temporal encoding there remains the design choice of whether to use oscillatory or one-shot dynamics. Both oscillatory and non-oscillatory versions work, but the oscillatory works a great deal better because it gives the competitive process a number of risetimes in which to operate, if necessary. By choosing time constants properly, the competitive result of one cycle can be carried over into the next, so that if one *rm* has started to prevail over the others in one cycle, it will be more effective in blocking the inhibition of stage r to stage b gating by the associated *ri* in the next cycle. At the beginning of a new cycle, when the incident signals are just beginning to rise, a small differentiation in inhibition is far more effective than when the incident signals are nearing or have already reached saturation. Because of the temporal encoding described above, the smaller contributors to the aggregate signal will not even have begun to rise by the time the stronger earlier signals have established an inhibitory obstacle via *ri*. Thus oscillatory dynamics in conjunction with temporal encoding contribute to the virtuous cycle. The differentiation in *rm* responses increases with each cycle.

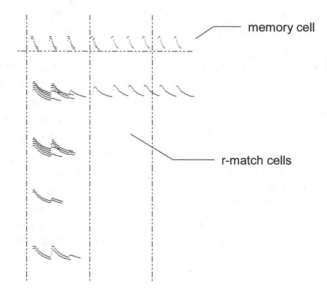

Fig. 5-9: Oscillatory dynamics of memory and r-match cells.

At the extreme, in overdriven situations it may take a few cycles of periodicity competition to converge to the recognition state (see Fig. 6-10, Chapter 6). It is rare for the neuronal circuit to take more than five cycles to converge, and three is more normal. Fig. 5-10 and Fig. 5-10 illustrate how far the convergence toward recognition proceeds during each risetime of the signal. A great deal of computation therefore takes place during each pulse, far more than is envisaged in any quasi-digital or quasi-boolean interpretation of cellular computation.

The simultaneous selection of memory and mapping can be seen most clearly in the activity of the *ri* cells and stage-m forward memory cells. Fig. 5-10 shows the activities of a few of these cells in the CMOS implementation of the circuit. The input field to stage r forward contains two possible targets: one corresponds to a previously captured memory, the other doesn't. To complicate the issue, the activations of the non-target inputs have deliberately been made stronger than those of the true target inputs. Initially all the *ri*s are inactive, so all stage-r signals are gated onto stage b, and all memory cells respond. Their response activates the *ri*s which can be seen decaying after the initial response of the memory cells. At this point the ultimate winning *ri* is more strongly activated than the ultimate loser, which is the opposite of the desired order. This inversion is due to the stronger activation of the non-target input. But on the second cycle the responses of *ri*s reverse into their ultimate order. This indicates successful competition by the *rm* cell associated with the weaker *ri* cell.

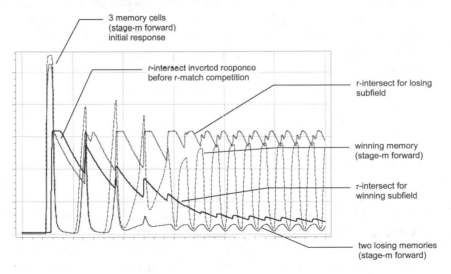

Fig. 5-10: Neuronal circuit dynamics (CMOS model).

The predominance of the ultimately winning *rm* cell results from the differentiation of the response of the memory cells, which is already taking shape by this second cycle. The destined losing memory cells are both lower in amplitude and lagging behind the destined winner. At the same time, the *ri* which controls (by inhibition) the gating of the target subfield onto stage r has fallen far below its non-target competitor. As a result the signals produced by the true target are becoming by far the largest component in the aggregate signal on stage b forward, and the differentiated response of the memory cells clearly shows this to be the case. By the fourth cycle the losing memory competitors are essentially non-responsive. By the sixth or seventh cycle the circuit has settled into its recognition state.

The oscillatory dynamics also provide a means of distinguishing state by periodicity. The circuit has three stable states, two of which were seen in Fig. 5-7: the recognition state denoted by a fast oscillation (optimistically labeled gamma), the non-recognition state denoted by a slow oscillation[2] (so far unnamed), and a silent state. The silent state only occurs when the input field is completely or nearly inactive.

One application of these distinct periodicities is an extension of the circuit that performs shifts of attention from an already recognized target in the input field to one or more secondary targets. The circuit extension detects the short periodicity of the recognition state and after a certain number of oscillations in that state it uses the matched pattern projected backward through the mapping to stage r in order to inhibit the matched elements of the target. This interrupts the recognition state and allows the circuit to hunt for another recognition; this continues until the input field has been stripped of all elements which match patterns in memory. Then the circuit is left in the slow oscillation, non-recognition state or entirely quiescent. The final slow oscillation state can be clearly seen in the 3D activity plot of Fig. 6-11, Chapter 6, after the faster oscillations of the two successful attention shifts.

5.6 Collusion

As with the algorithmic circuit there are conditions in which the neuronal circuit behaves less than ideally. These arise in collusion conditions in which spurious combinations of a few mapped input subfields happen to evoke a strong, stable recognition from a memory or memories to which they do not individually correspond. In this situation the spurious mappings outcompete valid mappings for long enough to completely suppress the latter before competition finally breaks up the spurious combination. As a result, the circuit falls into a non-recognition state when in fact there are targets which could be recognized. With its non-linear dendritic combining function the neuronal circuit is more resistant to this mode of failure than the algorithmic circuits demonstrated in earlier chapters.

---

[2] Even with a fixed clock gating the input, the non-recognition periodicity will be a longer clock multiple than the recognition periodicity. This is due to longer inhibitory blocking of the input in the first stage.

5.7 Neuron Function: Idealized versus Biological

In the beginning of this chapter there was a caveat concerning a number of departures from biological reality in constructing the model. Are any of the differences between the idealized neurons employed in the model and biological neurons fatal to the realization of a biological map-seeking circuit? There are a number of obvious areas to consider.

The neuron model adopted here breaks up the dendrite into a set of regions whose outputs are combined in a tree of dyadic (two input) operations. The combining operations at these bifurcations may be linear or non-linear. The linear mode mimics a simplified passive dendrite, the non-linear mode interposes a threshold which makes the combining operation more *and*-like. The nearest biological relative appear to be active zones in some dendrites. The majority of combining operations in the models demonstrated in this book are additive, emulating passive dendrites.

The regions themselves consist of one or more synapses. If the region is excitatory then its output is zero or positive, even if it contains within it one or more inhibitory synapses. In excitatory regions the excitatory synapses compute a similarity function between their respective weights and inputs. In the simulation this is computed as a thresholded dot product of the weight and input vectors. How realistic is this? A good deal of modeling work has been done to establish whether neurons can compute a match function similar to a dot product. A model investigated by Blackwell, Vogel and Alkon[3] demonstrates exactly the behavior desired in a set of synapses that would constitute a region in this model. The similarity function computed in their model is sharper than a dot product because a single mismatch between "weight" and signal has a strong shunting effect on the resulting output. This is most critical in the *rm* regions, in which two inputs must be effectively multiplied, so the absence of either vetoes the region's output. The two-synapse model demonstrated by Blackwell, et al. exhibits exactly the desired behavior. Their results are in line with those of other experimenters, but their test conditions are particularly well suited to the dendritic model used here. While this modeling is not verified by testing on biological dendrites, it is based on well-established dynamics. So it provides as solid support as is currently available for the assumptions used in the region computations in the neuronal model.

Nature's reliance on spikes to transmit signals along the axon and into the synapse poses no obstacle, as suggested earlier. Instrumentation of real cells and compartmental models agree fairly well that the waveform of post syn-

---

[3] Blackwell, Vogl, Alkon 1998

aptic potentials in the dendrite induced by spike input is a reasonably smooth pulse with sloping rise and fall. This is quite similar to the waveforms which occur throughout the idealized and CMOS realizations reported here. In the neuronal model computation takes place inside the dendritic structure; in real dendrites signals also interact in their smooth pulse incarnation. The spike is an efficient way of transmitting the result once the computation has taken place, particularly if maintaining the fidelity of the timing of the signal is more critical than maintaining the fidelity of its amplitude, as it is in this model.

The relatively slow propagation in biological axons and dendrites is an important difference from the idealized and CMOS realizations. The magnitude of axonal delay is not the primary issue, for assuming a 1 meter per second propagation of axonal signals and a loop of several mm, the axonal delays amount to only a few milliseconds. On the other hand, dendritic delays in pyramidal cells can amount to 10-20 milliseconds, from the most distal synapses to the soma.[4] Therefore the delays in a single cell pair could exceed the 40 Hz gamma frequency so often associated with cortical function. This fact alone unambiguously disqualifies the cell-pair chain as the mechanism producing oscillation, as assumed in the model, and makes an external oscillator more likely,[5] as suggested at the beginning of this chapter. The frequency of such an external oscillator might be altered by the averaged strength of the backward signal, with a net effect very similar to the loop used in the circuits demonstrated here.

An equally important consequence is that the backward pulse cannot be coincident with the forward pulse that evoked it. In the *rm* dendritic structure forward and backward inputs are paired locally, each pair contributing to the correlation only if both inputs are present. But biological dendritic pulses have durations of only several milliseconds in the vicinity of the synapses which produced them.[6] Unless the circuit loop is extremely compact the backward signal would arrive too late to influence even the decay of the pulse which invoked it.

One possibility is a slower decaying state in the *rm* dendrite, triggered by the *rfwd* input, acting to effect the facilitation of the delayed *bbkwd* input. Such a delayed effect is common in biological neurons, but has generally been observed operating at the soma [7] which will not serve the purpose here.

---

[4] Agmon-Snir, Segev 1996
[5] Gray, McCormick 1996
[6] Douglas, Martin 1990
[7] Koch, Segev 2000

Whether the same or a similar mechanism produces a delayed conjunction effect near the synapses is not known to the author.

Another, more elegant possibility, is that the backward pulse is coincident with a later forward pulse at the *rm* synapses. From a computational point of view this would be equivalent to the action in the model, except that it would take an extra cycle or two for the *rm* to start taking effect. In a multilayer circuit, the forward and backward pulses have to be coincident at the *rm* of each layer within an interval approximately equal to the pulse width, unless there is delayed facilitation as just discussed. Even in the absence of delayed facilitation, if the delay per stage is close to a multiple of one half the pulse periodicity the forward and backward pulse trains can be phase locked. At 40 Hz the delay could be 12.5 or 25 ms: quite plausible in pyramidals. Weak excitatory connections from the forward to backward path can accelerate a lagging backward signal, and a weak excitatory connection from the backward to the forward path can accelerate a lagging forward signal.

A more difficult problem for the model arises from the variations in dendritic propagation due to the differences in synapse location. In *rm* cells decisions are local to the paired inputs, so pulse spread and facilitation at the soma would tend to override the consequences of this variation. However, in the *bfwd* dendrites the distance variation would have a damaging effect on the superposition, since the components from more distal synapses would be phase delayed and therefore treated as weaker. In biological cells this is further aggravated by the variation in synapses and their stochastic nature. The ordering property on which map-seeking relies requires strict monotonicity in the contributions of the components of a superposition. The accuracy needed to effectively implement that monotonicity can be relaxed by using highly orthogonalizing input encodings. But the degree of variation posed by propagation delays and synapse behavior would appear to defy whatever robustness could be achieved with encoding.

One solution, which biology may have lighted upon, is to use redundant paths for each signal. As will be seen in Chapter 7, the neural investment in map-seeking circuits need not be very high, relative to the neuron populations in the cortices. A redundancy of a half dozen or more is not prohibitive. If these redundant pathways make synapses on the target pyramidal at varying distances from the soma, so that the average electrotonic distances for all signals are about equal, the effect would be to spread and time-normalize the contribution of each signal. At the same time, the redundant connections would mitigate the variations in synapse behavior.

The single greatest improvement to the model would be to introduce dendrites with realistic propagation delays. Pulses would be shaped by synaptic and dendritic dynamics rather than the loop feedback of the current model. However, the computer resources needed to simulate such a model would be very great, for most cells in the model have thousands of synapses, requiring state parameters for each synapse rather than for each cell, as in the present model.

The scheme that was used in the current model was the least computationally burdensome way to guarantee the synchronization of the forward and backward pulses and thus confirm that with such synchronization the map-seeking dynamic is viable in an oscillatory neuronal circuit. It also proved the viability for analog electronic implementation before the CMOS simulation was undertaken.

Neuroanatomy: Where?

Because the mechanism being proposed here is general purpose, if it can be found at all, it will very likely be found in various forms in many locations in the visual system and elsewhere. A few candidates are suggested here. fMRI studies have confirmed that separated areas within the MT appear to be active during perception involving translation and rotation.[8] The IT may be another locus of map-seeking circuits, for cells responsive to disparity-defined curved surfaces have been identified there.[9] The motor and premotor cortices are very likely involved in computation of forward and inverse kinematics, as discussed in Chapter 4. Area 5 of the parietal is apparently involved in computing the sensorimotor transformations from the visual frame to the arm motor frame, as discussed in Chapter 7.

---

[8] Morrone, et al 2000
[9] Janssen, Vogels, Orban 2000

Chapter 6
Neuronal Circuit Behavior and Performance

6.1 Temporal Aspects of Circuit Performance

To this point the discussion has focussed on functional behavior of the circuits with little or no consideration of performance. This chapter presents a number of tests in which the temporal aspect is relevant. Because it is difficult to define a robust relationship between the rates of convergence of the algorithmic and neuronal circuits, any claims of biologically plausible performance must be supported by demonstrations of the neuronal circuit. Since the time constants of the model are not based on physical processes, temporal estimates are made here by counting oscillatory cycles. The assumption is that a certain amount of convergence is possible during a cycle because of the relative timings and shapes of the pulses, independent of time base. Therefore if a computation takes n cycles to converge, and those cycles are presumed to occur at 40-60 Hz, it takes $17n$ to $25n$ milliseconds to complete the computation.

The rationale for assuming that the progress toward convergence made in a cycle is roughly similar in the neuronal circuit and in biological cortices is indirect and needs some explanation. It is assumed that all interaction between signals takes place in local areas of the dendritic tree. The form of the pulse in the neuronal circuits is reasonably similar to the signal in the dendrite of a biological neuron near the point of origin, and the rise time is long compared to time needed to propagate signals locally in a biological dendrite. Therefore it is reasonable to assume that the interaction between signals arriving at several nearby biological synapses is dominated by their relative time of arrival and whether they are excitatory or inhibitory (whether shunting or otherwise). As seen in Chapter 5, these are the essential characteristics of the dynamics of the neuronal circuits.

Under the assumption of comparable computation per cycle, it will be demonstrated that the neuronal circuit operating at approximately 40 Hz is capable of performance consistent with biological visual systems. In its basic form the circuit is capable of recognizing and tracking moving targets amid clutter. Utilizing the oscillatory period to synchronize the frame changes, it is capable of identifying an updated location of the target pattern on every cycle. Also illustrated in this chapter are scene segmentation by attention shifting, repeated pattern identification and its use in recharacterization, and the use of regularity extraction to segregate stable percepts within a scene. All of these are performed in intervals consistent with biological vision.

All test cases employ a single layer circuit with 6000 cell pairs in stage r, 512 cell pairs in stage b, 40 cell pairs in stage m, and 3000 mappings. The input stage is equipped with auxiliary circuitry to effect attention shifts within the input field using the same approach introduced in Chapter 2. This works by suppressing areas containing recognized patterns after a certain number of oscillations and allowing the circuit to attempt to recognize patterns in the remaining unsuppressed areas of the input field.[1] For all the tests the only variation in the running conditions, other than input, is that the output of the attention focus subcircuit to stage r forward is inhibited for the tracking tests (though this is not strictly necessary because the target moves in an interval shorter than required to engage the attention shift).

6.2 Input Encoding

The advantages of encoding images with multiple feature sets have been discussed in Chapters 2 and 5. In these demonstrations, due to computer memory limitations, only two feature sets are used. For variety (and perhaps realism) the image sampling pattern employed in these tests is hexagonal.

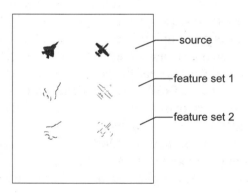

Fig. 6-1: Multiple feature filtering.

The input silhouettes are decomposed into two broadly tuned orientation fields by applying standard edge detection filters, Fig. 6-1. The resulting encoding of the search patterns is shown in Fig. 6-2.

---

[1] The attention shift circuitry is discussed at greater length in Appendix F.

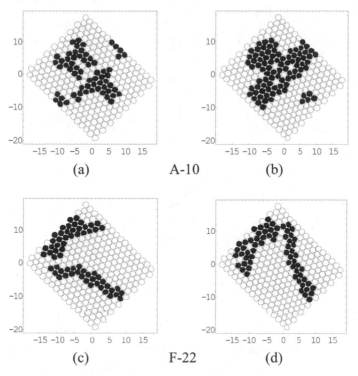

Fig. 6-2: Multiple feature encodings.

The effect of having two feature fields is to make the encoding somewhat more sparse. Since dimension of the memory and stage b fields is only 16 × 16 any increase in sparseness has significant benefits.

Two hexagonal samplings of the test image(s) are used as input to two extents of stage r, each representing a 50 by 60 feature field. Fig. 6-3 presents a schematic representation of a circuit with stage r partitioned into two retinotopically mapped extents, one for each set of feature inputs. The partition continues all the way to memory. To combine the two feature fields each *rm* (r-match) cell dendrite has two branches, one for each extent of stage r. The threshold of the node joining the two branches is adjusted to perform the equivalent of multiplication, so matches for both sets of features are required for the *rm* to become active. The same is true for *mfwd* memory cells. This makes the response of the *rm* and *mfwd* cells sharper than the dot product of the algorithmic circuit, and thereby increase the robustness and resistance to collusions, as discussed in earlier chapters.

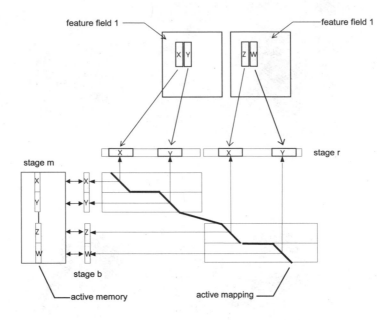

Fig. 6-3: Circuit organization for multiple feature encodings. Block schematic parallels layout of neuronal circuit of Fig. 5-5: mapping "fabric" bounded by stage r cell pairs on top and stage b cell pairs at left, stage m memory and its interconnections with stage b in rectangle to left of stage b. All stages in this circuit are partitioned into two feature fields, with multiplicative junctions in dendrites of *rm* and *mfwd* cells, as discussed in text. Bold diagonal line represents the winning mapping.

6.3 Tracking a Target Moving Amid Clutter

The continuous computation by neuronal versions of the circuit make tracking of moving targets inherent if the mappings allow. Fig. 6-4 shows a typical test of recognition of a target in motion amid clutter. Images of the two planes encoded in the dual feature representation are captured in two memory locations. To test tracking performance in a noisy environment the outline of the F-22 fighter jet is "accelerated" in a sequence of frames across a background cluttered with patterns of similar dimension and orientation to the target. The frames are changed every 120 simulation steps: 120 msec in "biological time," or approximately five oscillatory cycles. The inputs to stage r are the only externally controlled changes: no resets of the circuit accompany the changes of frame.

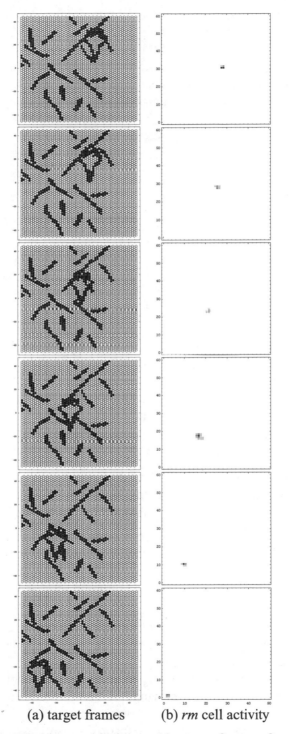

(a) target frames   (b) *rm* cell activity

Fig. 6-4: Tracking amid clutter. (a) target frames, frame 1 at top; (b) *rm* locating target, origin in lower left of memory field.

Note that in Fig. 6-5 (left) the two feature input fields have been superimposed to make the outline of the plane more evident. To the right of each input field is a snapshot of r-match activity evoked by that frame. It is evident that the r-match activity tracks the motion of the target, despite the clutter surrounding it. The sequence of convergences can be seen in the three dimensional plots of r-match activity, Fig. 6-5.

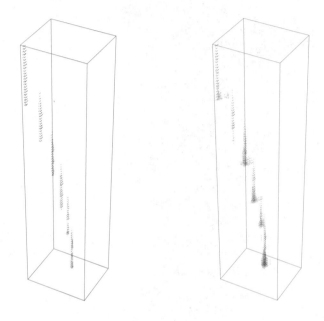

(a) test without clutter   (b) test with clutter

Fig. 6-5: R-match activity during tracking. Time is the vertical axis, increasing upward. Horizontal plane is the input field position of the mapping corresponding to the r-match cell. Amplitude of each r-match cell signal is shown by the length of the plot line toward the rear.

Fig. 6-5(a) shows r-match activity when the plane is moved across an empty background. The narrow vertical traces of mapping activity indicate that the correct mapping is selected almost immediately upon each frame change. Fig. 6-5(b) shows the r-match activity tracking the target against the cluttered background seen in Fig. 6-4. The broader clump of activity at the foot of vertical trace shows that a group of mappings clustered around the correct one become active after each change of frame, and it takes several oscillatory cycles to converge to the correct mapping. This is due to the background adding activation to the incorrect mappings and the competition circuit therefore taking a bit longer to suppress them.

Realism

Does presenting the target in a series of frames constitute a realistic test? If the visual cortex does function in a synchronized oscillatory manner, whether by the principles advanced here or some other principle, it in effect samples continuous inputs (or higher frequency inputs) from the LGN into frames at its oscillatory rate. The fact that motion picture frames are perceived as motion indiscriminable from real world continuous motion means that the mechanism is quite content with sequences of frames not necessarily synchronized to the cortices' own oscillation. The use of non-synchronized frames in the tests here demonstrates the circuit's capability in the more unnatural and more demanding of the two conditions.

Tracking versus Recognition

In the test above only two memories were trained. In pure tracking, the correct memory has already been identified and all others may be inhibited. This means that the input need only be mapped to a single memory, and this certainly improves performance. However, even with more than two memory candidates the tracking behavior remains viable. Fig. 6-6 shows the memory responses of a circuit trained with 24 rotations of the F-22 plus one rotation of the A-10. The input frames from the test above were used, but switched every 150 iterations instead of 120.

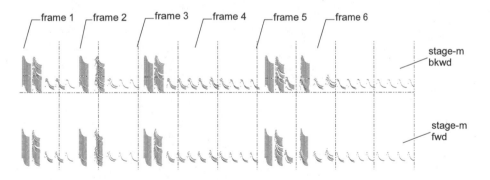

Fig. 6-6: Memory activity during tracking.

The memorized images are rotations at increments of 15 degrees, so as a result of the low resolution (and dilation of the memory pattern) the target pattern evokes some response from memories adjacent to the correct rotation, as can be seen in Fig. 6-6. Close inspection reveals differences in phase, reflecting the weaker activation of the incorrect memories. For each new frame, convergence to the correct plus adjacent memories normally takes place by the third cycle. For frame 4, where the switch happens to coincide with the activation of the memory cells, the convergence is sustained from the preceding interval. Frame 4 illustrates the advantage of synchronizing the frame switch with the oscillatory cycle of the circuit. Conversely for frame 5, where the synchronization is poor, convergence is partly accomplished by the third cycle, but fails in the next cycle, and finally resolves by the fifth cycle. Frame 6 is presented in synchrony with the fifth cycle, so convergence is preserved, as for frame 4.

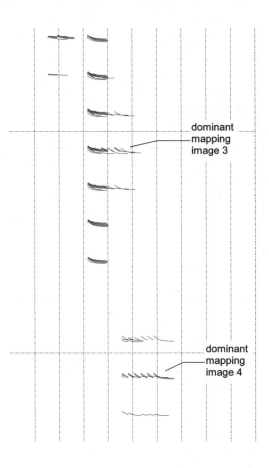

dominant
mapping
image 3

dominant
mapping
image 4

Fig. 6-7: R-match activity – unsynchronized and synchronized frames.

Examining the r-match activity, Fig. 6-7, from the same test also illustrates why it is highly advantageous to synchronize the frame switch to the oscillatory period. No time is spent resolving mappings between images 3 and 4 and between images 5 and 6 because the backward path is already active with the recognized pattern at the time the new image is presented. Thus the new mapping is immediately established, as can be clearly seen in Fig. 6-7. Compare the initial number of active mappings for image 3 versus image 4.

6.4 Attention Shift Tests

In Chapter 2, a sequence of recognitions within a single input image is demonstrated in Figs. 2-12 and 2-13. An autonomous version of the same action is implemented in the neuronal circuit. A simple extension of the circuit, driven from the backward path, inhibits the inputs to stage r which correspond to recognized targets so that the circuit can shift attention to other targets in the input field. A test image, Fig. 6-8(a,b), containing two different plane targets, is used to demonstrate the behavior of the circuit with its attention shifting capability enabled.

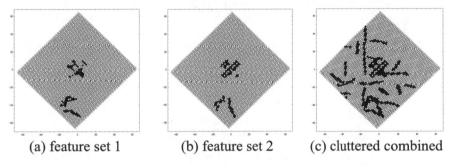

| (a) feature set 1 | (b) feature set 2 | (c) cluttered combined |

Fig. 6-8: F-22 and A-10 test pattern.

Memory is loaded with 24 different rotations (at 15-degree intervals) of the F-22 and one view of the A-10. The resulting circuit activity for the uncluttered input image is shown in Fig. 6-9. The two traces in Fig. 6-9 are stage-m memory backward and forward cell activity respectively. The first target recognized is correct rotation of the F-22. After a half dozen cycles of recognition state, the circuit automatically suppresses the F-22 target inputs by using the F-22 pattern projected onto stage r backward to inhibit the inputs to stage r forward. Then the circuit recognizes the remaining pattern: the A-10. After the second target is recognized and suppressed there are no inputs left to activate the circuit.

Further detail on neuronal attention shift circuitry appears in Appendix F.

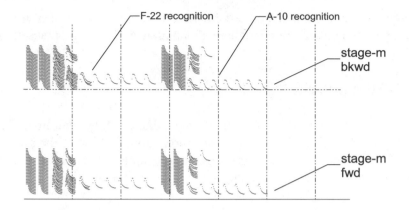

Fig. 6-9: Memory activity during attention shift. Memory patterns for 24 rotations of F-22 and one view of A-10. Single test image containing both F-22 and A-10, without clutter.

6.5 Scene Segmentation

The attention shift subcircuit in effect segments the input image into recognized components and an unrecognized remainder. The segmentation behavior depends on the activity patterns mapped back to stage r backward by the winning mappings. Fig. 6-10 shows stage-r backward activity for one feature component during an F-22 to A-10 attention shift. The input field is the cluttered version of the two plane image, Fig. 6-8(c). The 3D plot data has been thresholded at near peak value to clarify the location of the cycles.

The data of Fig. 6-10 are the intersection (in the set sense) of the *rfwd* activity and the *bbkwd* activity mapped onto *rbkwd*. We saw this intersection as an analytical tool in Chapter 2. Here it is a functional characteristic of the circuit. It is implemented by adjusting the threshold of the *rbkwd* cell so that the excitatory connection from the paired *rfwd* cell must be active for the *rbkwd* cell to respond to the backward mapped signal.

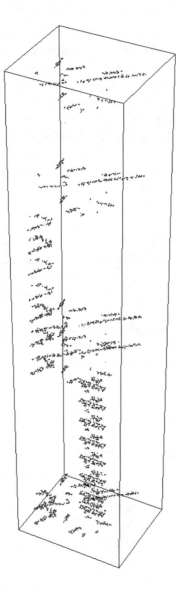

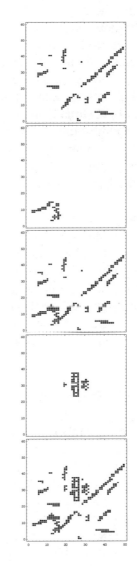

(a) oscillatory peaks, time on y axis   (b) snapshots

Fig. 6-10: Stage r intersection activity during attention shift for a single feature set. (a) Data at oscillatory peaks, time increases on vertical axis; (b) Snapshots of data during first cycle, A-10 recognition, first attention shift, F-22 recognition, and final non-recognition state. First cycle in bottom panel, last cycle in top panel.

During the first cycle in Fig. 6-10, most of the forward pattern appears in the intersection because all of the *ri* cells are active, gating signals onto all *rbkwd* cells. But by the next cycle the mapping corresponding to the A-10 is already at work, and it restricts all *rbkwd* activity to the subfield containing the A-10. After about eight cycles the A-10 input is inhibited by the activity on *rbkwd* driving the attention shift circuitry. Now the remaining input field, containing the F-22 and the noise, gets the attention of the circuit. Again all the mappings are active so the remaining *rfwd* pattern appears in the intersection. But after two cycles of this, the mapping for the F-22 takes control, and for five cycles the F-22 is the only pattern active on *rbkwd*. Finally the F-22 pattern on *rbkwd* drives the attention shift circuitry to inhibit the F-22 inputs. Now all that is left is the noise in the input field. But no pattern in memory matches this, so the circuit enters the slow non-recognition state oscillation, as evidenced by the long distance between the final state oscillatory peaks in Fig. 6-10(a).

By comparing the cycles of Fig. 6-10 to those of Fig. 6-9 it becomes evident that adding noise to the input field both reverses the order of recognition of the two planes and reduces the number of cycles needed to resolve the recognition of each. Apparently some component of the noise adds enough activation to the A-10 pattern to give it the edge initially. There is also enough additional *rm* cell activity from partially matched noise components to increase the inhibition in the *rm* competition. Increasing the inhibition in the *rm* competition hastens the resolution of the correct mapping.

6.6 Repeated Pattern Recognition

Another inherent behavior of the circuit is to concurrently locate repetitions of a recognized pattern. Since the competition between mappings in the circuit used in these tests was not tuned to be aggressive, multiple instances of the target in the input field are located by concurrently active r-match cells. An example of such behavior is shown in Fig. 6-11. Visual repetition is a powerful, usually pleasing, stimulus, whether geometrically ordered as in architecture, or statistically ordered as in such natural organizations as schools of fish or fields of flowers. In either case the process starts with the concurrent identification of the locus of many instances of the same or very similar targets.

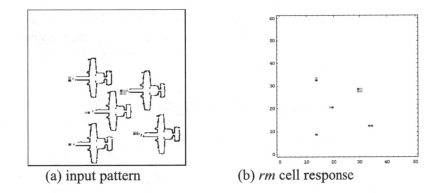

(a) input pattern   (b) *rm* cell response

Fig. 6-11: Repeated pattern location. (a) 5 A-10 outlines with location pattern of final active *rm* cells superimposed; (b) final *rm* cell activations, origins displaced from corner to airplane nose in memory field.

The capability of concurrent recognition of repeated patterns combined with the attention shifting behavior illustrated earlier gives rise to *visual popout* behavior. If a lone target pattern is located in the midst of a repeated distractor pattern, the target pattern will be identified in an interval independent of the number of instances of the repeated pattern. This arises because the repeated pattern creates very strong activation on stage b, so its pattern is recognized or captured first. But then the attention shift mechanism suppresses *all* instances of that repeated pattern because there is an active mapping for each instance, so the distractor pattern is mapped backward at the same time to all the locations in stage r in which it occurs. Using these backward signals the attention shift circuitry suppresses all the input instances. The leaves the solitary target pattern free to take the circuit's attention. In primates the time to identify a popout pattern is independent of the number of distractor patterns.[2] This is taken as general evidence of the parallelism of the processing of the input field. The mechanism proposed here is more a form of concurrency than parallelism. The only parallel resources engaged in repeated pattern detection are the translational mappings between stage r and stage b.

The circuit used in this demonstration is a single layer with only translational mappings. The psychophysics of repeated pattern recognition, however, provides some evidence that the human visual system may in fact deploy translation in one layer and rotation in a deeper layer, as do most of the algorithmic circuits used for demonstrations in this book. Identical targets with common orientation are more readily seen as repeating patterns

---

[2] Palmer, p 554-5

than identical targets with different orientations. An example of this is seen in Fig. 6-12(a). This is what would be expected if a single mapping in the rotation layer can match all the translations of the repeated pattern.

6.7 Recharacterization with Repeated Patterns

An implication of this repetition locating behavior is that a cascade of two of these circuits may recognize image targets rendered in different textures or components. This is another application of *recharacterization*, or using evoked mappings as data for subsequent stages of processing. The first step in repeated pattern recharacterization is to capture a sample of the elements in which a target image is rendered and then locate all instances of that sample in the input field. This form of recharacterization is an instance of *rendering transformation* or *feature transformation*. This use of recharacterization permits an object to be recognized across a wide range of renderings.

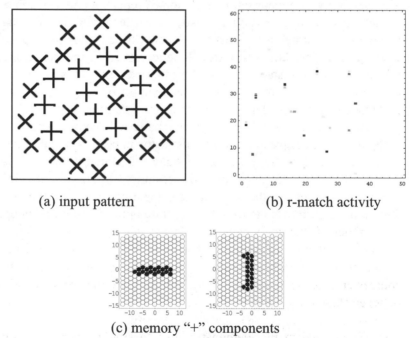

(a) input pattern   (b) r-match activity

(c) memory "+" components

Fig. 6-12: Recharacterization by repeated pattern. (a) Input global pattern ("a" on side) rendered by local "+" patterns; (b) r-match cell activity locating all instances of the memorized "+" pattern, offset by distance to origin in lower left corner of memory field; (c) two components of the memorized "+" pattern.

In Fig. 6-12(a) the "a" on its side rendered by the repeated "+" pattern is detected by the circuit as a pattern of concurrent r-match activity, Fig. 6-12(b). The circuit in effect *abstracts* the large pattern produced by the repetition of the rendering pattern ("+"). It is the rendered abstract pattern that is of primary interest, so the output of the r-match population is used as the input field to a second tier circuit which captures and/or recognizes the abstraction of the pattern recharacterized by the first tier circuit. The next time the "a" on its side is presented, though it may be rendered by a different repeated pattern, the second tier circuit will see the same pattern of r-match activation from the first tier circuit.

The behavior demonstrated in Fig. 6-12 mimics an often demonstrated human psychophysical visual capability.[3] It also arises in experiments which establish the precedence of "global" versus "local" processing.[4] In these experiments the global figure is the abstract or recharacterized pattern, while the local figure is the characterizing pattern. Fig. 6-13 is a block diagram showing the relationship of the two tiers of a highly abstracted recharacterization circuit.

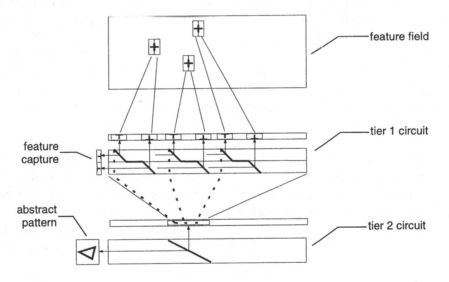

Fig. 6-13: Recharacterization in two-tier circuit. The mappings of the tier 1 circuit are the input field to the tier 2 circuit. An input pattern rendered by some repeated feature activates the tier 1 mappings corresponding to the location of that feature. The tier 2 circuit recognizes the abstract pattern of the locations identified by tier 1.

[3] Marr, p 94-5, Figs. 2-34, 2-35, and p 192, Fig. 3-46a

[4] Palmer, p 356-7

6.8 Regularity Extraction in the Learning Process

Though we have seen the effectiveness of the circuit in recognizing a memorized pattern amid noise and clutter, so far the patterns which have been used to train the memories have been presented in a noiseless environment. This is not a realistic condition for a mechanism that has to capture its memorized percepts from most natural environments. On the initial presentation of a new target there is no way to distinguish what parts of the image constitute the target and which the background, noise or distractors. However, on repeated presentation of the same target in the presence of differing backgrounds, noise, and distractors, the part of the image that constitutes the target can be segregated if it remains constant, or undergoes transformations which are accommodated by the circuit. The process which performs this segregation is termed *regularity extraction by common subset identification.* It is another inherent behavior of the circuit: a consequence of the backward path through the mappings.

This process is made possible because the signal pattern projected onto the backward path is the memorized pattern, which may be different to some degree from the signal pattern on the forward path which evokes it. Multiplying the two signals yields a set of intersection results, as we first saw in Chapter 2. This intersection represents the stable part of the image. If a target has moved relative to the background, there will be two stable intersections with different translations: the target and the background. The mapped extent of each is the basis for isolating the stable percept in one or a few iterations.

The circuitry to accomplish this is quite simple. The intersection is performed by multiplicative or gating connections from *rfwd* to *rbkwd* and similarly in the other direction. This restricts *rbkwd* activity to the subset common to both signal patterns, and this signal is then used to gate *rfwd*. The details of one implementation and some discussion of the technicalities can be found in Appendix G. The result of this process is shown in Fig. 6-14.

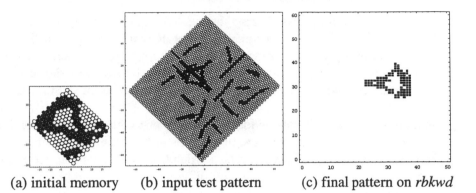

    (a) initial memory   (b) input test pattern   (c) final pattern on *rbkwd*

Fig. 6-14: Regularity extraction.

Fig. 6-14(a) is the initial pattern captured in memory: the F-22 in the presence of a "noisy" background (derived from earlier tests). The input pattern, Fig. 6-14(b), is the F-22 displaced against the same noise pattern. Fig. 6-14(c) shows *rbkwd* activity after stable recognition is established. The result is the intersection of the captured pattern and the section of the test pattern located by the F-22 target. If this *common subset* pattern is compared with the isolated F-22 training patterns presented earlier it will be seen that the original pattern has been effectively isolated from the noise of both training pattern and test pattern.

The displacement of the F-22 in the test pattern of Fig. 6-14(b) is a fortunate one, in that the intersection of the memorized and test patterns yields almost exactly the original F-22 image. This is not always, or usually, the case. More normally some part of the noise will remain. As a consequence the process of regularity extraction by common subset identification is usually a multi-step process. The result of the first step is captured, replacing the noisier version, and is then used to recognize the target again when the background changes. At each step more of the noise is removed from the memorized percept until the stable subset is isolated.

It should be apparent, though not demonstrated here, that since the intersection process takes place in stage r, upstream of the mappings, the target can be any implemented transformation of the memorized image. Therefore either the motion of the observer or the target can produce the shift necessary to extract the common subset corresponding to the stable image of the target. This process might be called *identity from motion*, and the virtual motion of stereo vision should be adequate to launch the process.

The process by which the first percept in the sequence is selected has not been addressed. Some attentional mechanism has to activate the mapping or cluster of mappings, which map the area containing the target onto stage-b to capture it. This requires some means of deciding what part of the input field contains a candidate target. It could be bootstrapped by using a previously captured percept component or inherent feature to steer the mapping. Or it could use motion, a color or some texture feature as an attentional cue.

Is there any psychophysical evidence that this procedure is used by biological visual systems? Humans are able to distinguish the outline of a set of random dots moving relative to a background of dots so long as the motion continues. When the relative motion of subject and background stops, the outline vanishes. The process described above explains this ability. An apparent manifestation of this capability in another domain was reported by J. Saffran et al[5]. Eight-months-old infants appear to be able to identify sequences of phonemes corresponding to synthetic words without any cues of the boundaries other than the higher probability of phoneme sequences corresponding to words, and the lower probability of phoneme sequences corresponding to non-words. A two minute training period was sufficient to allow the infant to perform this discrimination in a test sequence. Saffran attributed the word/non-word discrimination to an inherent ability to differentiate the statistics of the phoneme sequences during training. However, the author was able to duplicate the results reported by Saffran by applying the stable subset identification procedure described above to training sequences corresponding to those used in the experiment with infants. (Author's results unpublished.)

---

[5] Saffran, Aslin, Newport 1996

Chapter 7

Neurophysiological, Psychophysical and Other Evidence Seen Through the Theory

7.1 Purpose

A theory is distinguished from a model by its ability to provide a unitary explanation for a variety of phenomena. In this chapter a number of neurophysiological and psychophysical results will be visited and analyzed using the map-seeking circuit as the core of the explanation. Most of the psychophysical phenomena have been largely without explanation up to now, and certainly not unified by an explicit mechanism.

7.2 Physiological Evidence

Visual Cortex – Evidence of Stage r Intersection Signal

Though the body of neurophysiological evidence that bears directly on this theory is limited, a number of recent experiments relying on cell recordings in V1 and V2 have provided evidence of "top down" or recurrent inputs in primary visual cortex processing. The results of two groups, intended to address the existence of recurrent inputs to early visual cortices, also provide evidence of far more specific organization and dynamics remarkably consistent with aspects of the neuronal circuit.

Supèr, Spekreijse and Lamme[1] report a contextual modulation of response of V1 which differentiates figure from ground when the figure is "seen" as opposed to when it is "not seen." The modulation consists of a suppression of the cells with receptive fields in the background areas relative to those in the foreground areas 100-120 msec after the presentation of the stimulus. The differentiation is only present when the animal (an awake monkey) indicates having processed the presence of the foreground figure by launching a saccade to the figure. The figure used is a filled square formed by lines of orientation different from the lines forming the background, Fig. 7-1(a). The cells on the boundary of the figure exhibit the strongest response, but the cells with receptive fields (RFs) in the figure interior exhibit responses well above the background, Fig. 7-1(b).

---

[1] Supèr, Spekreijse, Lamme 2001

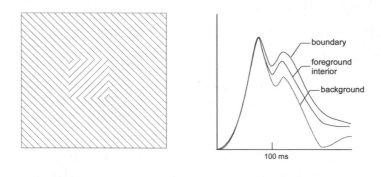

(a) figure-ground display   (b) V1 cell responses

Fig. 7-1: Results from figure-ground experiments of Supèr, Spekreijse and Lamme (simplified from original figures). (a) test display
with figure and ground distinguished by orientation of texture; (b)
responses of V1 cells with RFs on boundary, foreground interior
and background when figure is seen – data merged from multiple
graphs.

These results correspond remarkably with the stage r intersection data seen
in earlier chapters, in particular Fig. 2-12 and Fig. 6-10. Interpreted using
the neuronal map-seeking circuit as a model, the data provide a rather detailed picture of the interaction of the forward and backward signals.

Two backward signals appear to be converging on the V1 cells: a signal corresponding to a memory of a square boundary and a signal corresponding to
a memory of the interior texture of the square. Initially, at about 50 msec,
there is a large response in which none of the areas of the input pattern are
distinguished. This corresponds to the first cycle of Fig. 6-10. Interpreted in
map-seeking terms, this is before the circuit starts converging, so any response from memory is mapped across the entire *rbkwd*. During that
interval the intersection signal is identical to the *rfwd* signal. That is to say,
no differentiation has taken place.

About 50 msec later the responses have differentiated. The boundary signal
now evokes a stronger intersection signal with the forward input pattern than
the interior texture signal, and the responses of the V1 cells with RFs in the
background area are significantly lower. Supèr, et al. refer to this difference
as "modulation." The interval from 120 to about 200 msec, during which all
responses decline, corresponds to the second and third cycle in Fig. 6-10.
This is the period in which the mappings are being resolved and the signal
on the backward path is being pruned down to the matching memory pattern. When the activity of the cells in the background reach zero and only

142

the foreground cells are active, the state corresponds to the fully converged first recognition in Fig. 6-10. No backward signal is arriving at the background cells of V1, and consequently the intersection signal is zero. However, the foreground cells are still receiving backward signals corresponding to the correct mapping of the memory of the square, and the intersection signal for these is non-zero.

If the two initial peaks in Fig. 7-1(b), separated by approximately 50 msec, indicate a periodicity in the process, it takes about four of those intervals for the background activity to reach zero. This is roughly consistent with the behavior of the neuronal circuit as seen in Fig. 6-10(a), which by the fourth cycle has completely suppressed the background. 50 msec is a longer periodicity than the 40 to 60 Hz (25 – 17 msec) normally associated with the visual system, but in some tunings of neuronal map-seeking circuits the initial cycles of convergence are longer than those later in the convergence. This can be seen in Fig. 5-9.

The data of Fig. 7-1(b) has another implication. The average response of the cells within the target decreases once differentiation or modulation has been established. This is not observed in map-seeking implementations because the target pattern signals and the memory pattern signals are at saturation. So even the superposition on the backward path prior to full convergence cannot make the intersection signal greater than unity. But apparently this is not the case in biological cortices. The converged-state response of the cells is significantly lower than their maximum. One explanation is that the cells normally operate well below saturation so a superposition input produces greater activity than a single input. An alternative explanation is that general inhibition increases during the process of recognition. We will see further evidence of both interpretations.

The sequence of events described above applies when the monkey has indicated having seen the figure by launching a saccade toward it. When the foreground figure is not seen by the monkey, as indicated by a failure to execute a saccade to the target, both foreground and background remain equally active at somewhat lower levels than the foreground in the "seen" trials. Again, this is consistent with the undifferentiated response across stage r when there is no recognition, and therefore no convergence to a single mapping. In this case the signals from memory are mapped across the entire backward path and the intersection pattern is consequently identical to the entire forward signal, as seen in the last cycles of Fig. 6-10. The slower non-recognition periodicity might account for the lower average V1 responses in the "not seen" trials. In the "not seen" trials average response of the cells also decreases, unlike the neuronal map-seeking models. Since there should be no reduction in the number of components in the backward superposition, this again suggests an increase in general inhibition during the process.

Supèr, Spekreijse and Lamme interpret their results to implicate the involvement of a top-down or backward signal corresponding to the foreground figure. Interpreting their results using the far more explicit model of the map-seeking neuronal circuit supports their conclusion, and provides a deeper glimpse of the mechanism at work.

Some results reported by Lee and Nguyen[2] enrich this picture. In an experiment designed to determine how virtual contours are processed, recordings were made from V1 and V2 cells located at the virtual and actual boundaries formed by a Kanizsa figure stimulus, Fig. 7-2(a) left. The monkeys in the experiment are trained with outlined squares, Fig. 7-2(a) right, of the same dimensions as the virtual square to be produced by the Kanizsa figure.[3] When the monkey is presented with the outlined square display, cells with RFs anywhere along the boundary (Fig. 7-2(a) right, location 2) respond strongly 50 msec after presentation and sustain their response at a lower level after 100 msec. When presented with the Kanizsa figure the V1 cells with RFs on the virtual contours (Fig. 7-2(a) left, location 2) produce an attenuated response at 100 msec which does not sustain. The responses are seen in Fig. 7-2(b). The Kanizsa figure also produces responses in the cells along the boundaries inside the "pac-men" (Fig. 7-2(a) left, location 1) like those produced by the outlined square.

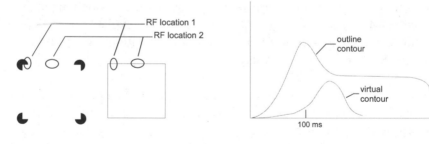

(a) test displays, recording locations   (b) V1 cell response, location 2

Fig. 7-2: Results from virtual contour experiments of Lee and Nguyen (simplified from original figures). (a) test display with Kanizsa figure, left, and outline training and test pattern, right – location of V1 cell RFs shown schematically; (b) responses of V1 cells with RFs in location 2 for outline and Kanizsa test displays.

---

[2] Lee, Nguyen 2001

[3] The use of the outline square as the training pattern was reported by personal communication.

To interpret this evidence from the map-seeking circuit perspective one must visualize the signal corresponding to the learned pattern, the outlined square, mapped onto *rbkwd* aligned with the non-virtual corners in the input field created by the pac-men.[4] Where the two signals are both present the strong and sustained response of the intersection signal is expected. What is not expected in the map-seeking circuit interpretation is the transient response corresponding to the virtual contour seen in Fig. 7-2(b). One does expect a delayed backward signal in the virtual contour locations since the learned figure was complete. But in multiplying that backward signal with null areas in the forward signal one would expect a null intersection signal, the observed final result, Fig. 7-2(b). These dynamics suggest that the intersection is implemented by something other than a pure multiplication. Rather, the responses to both displays are consistent with a gating function more similar to thresholded addition acting in the presence of a generally rising inhibition.

The timing and responses of the V1 cells reported by Lee and Nguyen are remarkably consistent with those reported by Supèr, Spekreijse and Lamme. The virtual boundary signals Lee and Nguyen attribute to V2 cells show a slightly earlier rise and a much slower decay. The earlier rise is to be expected of signals returning along a backward path. The slower decay suggests the fast-acting general inhibition is localized in V1.

Other published reports show similar evidence of backward mapped gating signals to V2 and V1,[5, 6] and are similarly explained by the map-seeking mechanism.

Visual Cortex – Evidence of Phase Coded Superpositions

There is some neurophysiological evidence for the encoding of signal magnitudes using phase. Victor[7] reports the results of a series of experiments to determine how spike timing carries information in V1. A number of metrics were designed to distinguish whether information is carried in the length of the intervals between spikes or in the absolute time of a spike after stimulus onset (or a transient marking it.) For a number of stimulus types the evidence strongly supports absolute spike timing as the primary encoding. Additional experiments on pairs of cells indicate that the identity of the cell

---

[4] See discussion of virtual contour response later in this chapter.
[5] Pascual-Leone, Walsh 2001
[6] Zhou, Friedman, von der Heydt 2000
[7] Victor 2000

upstream is also significant in distinguishing the information content carried by members of the pair. This was interpreted to indicate that the pairs were not carrying redundant signals. For contrast signals the timing resolution was determined to be 10 msec or less.

The methodology and data reported by Victor make a good case for a biological version of the kind of temporal or phase coding utilized in the neuronal circuit. This is particularly supported by the independence of the signals carried by the observed pairs of cells.

An additional set of observations reported by Victor provide evidence that is consistent with the superposition principle. In the process described in Chapter 5 one would expect that the information content of a signal from a cell that carries a superposition to increase as the convergence silences contributors to the superposition (if the cell carries a surviving signal). An information gain over time is observed in a series of tests reported by Victor. The gain in information content varies with the pair tested, but individual cells in the two pairs reported show average gains from the 100 msec interval to the 256 msec interval (both starting at onset), indicating that a gain in information content occurs after 100 msec. This result casts a different light on the interpretation of the general drop in activity after 100 msec seen in Fig. 7-1(b) as purely a manifestation of increasing inhibition. There must also be a reduction in inputs to the cell that are not correlated with the target pattern. This is consistent with the map-seeking mechanism.

Since the cells tested are in V1, the signals recorded most likely show the influence of backward path inputs, as reported by Supèr, et al. and Lee and Nguyen. In the simplest map-seeking interpretation of these results the pruning will have taken place on that backward signal, not on the forward signal into V1. (Though the forward signal into V1 might also reflect a backward path influence through the LGN.) Though the observed behavior is consistent with map-seeking it is not conclusive, since there are other mechanisms which would exhibit similar gains in information content during convergence.

Motor and Premotor Cortex – Evidence of Projection of Visual Circuit into Kinematic Circuit

In Chapter 4 map-seeking circuits are shown to be capable of solving the forward and inverse kinematics of multi-segment limbs. Kakei, Hoffman and Strick, cited in that chapter, report recordings from cells in the ventral premotor cortex and motor cortex which suggest distinct populations of cells encoding target position and joint posture. Later in this chapter a psychophysical behavior known as the kinetic depth effect is suggested to be a by-product of the sensorimotor transformation mechanism necessary to interpret views of limb posture into the kinematic frame. The proposed mechanism uses a sequence of map-seeking circuits to project a recharacterized visual frame representation of the limb segments into the body-centered frame of the stage r fields of the limb kinematics circuitry.

Graziano, Cook and Taylor[8] report recordings in monkeys from cells in area 5 of the parietal lobe which both encode the posture of a limb hidden from view and the posture of an artificial limb in the monkey's view. The sensorimotor transformations from the viewplane representation of a limb to limb posture encoded in a body-centered frame, and finally encoded in a set of joint centered frames, requires applying the constraints of limb geometry and joint range. This is readily expressed as a mapping problem, part forward and part inverse, as will be seen later in this chapter. The discovery of cells that manifest responses to visual frame representations of the limb, and manifest responses to corresponding joint postures, confirms that visual frame signals are projected into the limb kinematics frame. Just such an arrangement of map-seeking circuits is discussed later in this chapter because it explains a number of visual behaviors, as well as explaining our remarkable facility to mimic observed limb postures.

7.3 Psychophysical Evidence

A range of visual psychophysical behaviors can be provided with quite simple explanations based on the functional capability of a map-discovery mechanism. A number of these will be discussed. There is physiological evidence that many of these behaviors arise in different cortical areas, and by their natures the behaviors imply different resolutions, use of different attributes in the feature vector, and different repertoires of mapping. The fact that map-discovery provides a parsimonious and natural explanation for these diverse behaviors argues for the involvement of variants of the same circuitry in the responsible cortical areas.

Stereopsis and the Julesz Effect

As discussed earlier, two map-seeking circuits with translation mappings, and possibly small ranges of perspective and scaling, can be arranged to produce disparity signals for both vergence and depth computations. The relationship of the two circuits is shown in Fig. 7-3. A single central field (from the dominant eye, for example) is projected along the backward path

---

[8] Graziano, Cooke, Taylor 2000

to stage-r extents for each eye. A mapping in each circuit will be aligned on the common subset of both images *whether or not the eyes have properly converged.*

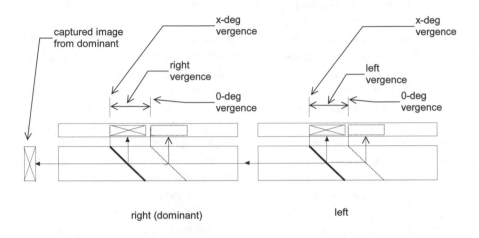

Fig. 7-3: Stereo vision architecture block diagram. Block schematic conventions as in Fig. 6-3. In this circuit the separate partitions of stage r, corresponding to each eye, have independent mappings, with separate competition circuits, so two mappings are found to a common memory pattern, one for each eye.

If either or both eyes are improperly aligned the active mappings will not correspond geometrically, and thus constitute a direct error signal from which convergence may be adjusted. This implies that the convergence need not be perfect even at its endpoint, since translation capability in the circuits for both eyes will tolerate a misalignment. Even if the left and right image inputs are physically misaligned, once the mappings for both are established, the signals from both eyes can be intersected in the stage-b forward path (or in the stage-b backward path if the intersection operation described above is in action) to effectively isolate the target (the common subset) at the converged distance from the parts of the image nearer or further away. It should be apparent that this process does not depend on the image having any recognizable aspect, but simply the match of the largest common subset within the last captured "frame." This mechanism fully explains the effect reported by Bela Julesz[9] wherein two planes in a random-dot stereogram are recognized from their synthetic disparities. It also ap-

---

[9] Julesz 1960

148

pears that some recharacterization must precede this circuitry to explain the ability to extract depth information from differently rendered images in each eye.

Barber Pole Illusion – No Aperture Problem, and Beyond

The detection of motion has been generally thought to precede the identification of an entire percept, and as a result a number of visual behaviors have either no explanation or rather tortured ones. An example is the familiar barber pole illusion. The old-time spinning barber pole with its helical red stripe is perceived as having upward velocity, though in fact the rotation imparts only horizontal motion to the surface of the pole. For a visual system that computes motion from the local velocity of image discontinuities the barber pole poses a problem, since observed locally the direction of motion is ambiguous. Some explanations of the apparent upward motion resort to the corners of the stripes, but if the barber pole were emerging from a cloud so that there were no distinct corners, the illusion would not be interrupted.

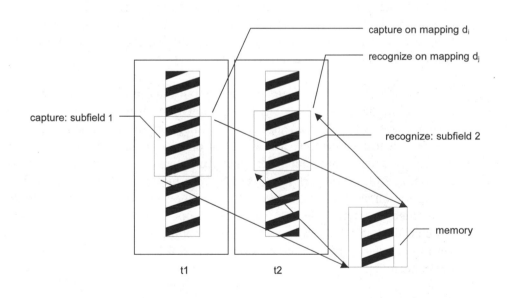

Fig. 7-4: Barber pole illusion. Two frames of the barber pole pattern, at times t1 and t2, correspond to a horizontal rotation of the pole. The pattern captured in memory, at right, from t1 matches the pole pattern at t2 with a translation in the vertical direction.

Under the theory presented here the perception of vertical motion is very simply explained. A circuit captures an image of the barber pole large enough to contain its full width, or at least a vertical boundary and part of the stripe pattern. See Fig. 7-4. On the next oscillatory cycle the rotation of the pole has caused the pattern of stripes on the input field to be translated vertically by a small amount. The mapping corresponding to that vertical translation is the next one activated. The impression of vertical motion is produced either by a sequence of activations of mappings corresponding to vertical translation or by the same sequence of mappings providing a velocity signal to launch smooth pursuit of the apparent vertical motion. This is precisely the same tracking behavior demonstrated in Chapter 6.

Barber Pole in Depth

A more subtle variant of the barber pole illusion requires consideration of the role of boundary ownership in the processing. When observed through an aperture whose disparity places it nearer the viewer, the barber pole illusion is weakened or lost. Yet when the plane of the pole pattern has disparity that places it nearer the viewer, the illusion remains intact. This dichotomy has been taken as evidence for a simple boundary ownership heuristic: a shared boundary is "owned" by the nearest object to the viewer which could have produced the boundary.

This boundary ownership heuristic can be translated into an explicit mechanism within the map-seeking circuit interpretation. It has been suggested in earlier chapters that biologically realistic image representations most likely utilize rich feature vectors, including not only boundary attributes, but numerous others including depth (distance from viewer). A depth attribute can be computed from disparity or parallax from motion or hypothesized from configuration cues. In Chapter 4 it was demonstrated that a scene can be segmented by intersecting the backward mapped signal with the input field to determine the effective "extent" of the mapping. This can be done as well with the boundary of a figure as its interior, and is one way of filling in the depth attribute.

Returning to the barber pole in Fig. 7-4, it can be seen that without inclusion of a border in the captured pattern, the mapping of the captured pattern alone could "slip around" the interior of the striped area of the input field, and thus the simple upward progression of mapping locations would be lost in a multitude of mappings. Therefore a vertical external boundary is necessary to enforce the proper sequence of mappings. However, the failure of the barber-pole-through-an-aperture illusion tells us that the boundary must share depth (to some degree of accuracy) with the striped plane to be included in the mapped extent. In the mechanism of a map-seeking circuit this implies that not only the boundary attribute but the depth attribute of the feature vectors of both input field and memory are part of the comparison performed by the r-match circuitry.

The inclusion of the boundary depth attribute in the mapping match is only the beginning of a discussion of richer mapping behaviors made possible by using more attributes in the feature vector. An endstop attribute, denoting the termination or abrupt bend of a line or boundary, has many uses, one of which will be seen in the kinetic depth effect discussion below. In the context of the current problem, an endstop attribute is a strong cue for hypothesizing a value of a depth attribute when the latter cannot be determined from direct inputs such as disparity. An endstop co-located with a transverse edge indicates a likelihood that the endstop, and whatever is attached to it, sits at a greater depth. An endstop co-located with another endstop indicates a likelihood that the two sit at the same depth. Relative depth attributes can therefore be assigned by local decoding of the presence or absence of endstop and/or boundary attributes in the same or adjacent feature vectors. It should be noted that this decoding "logic" is not part of the map-seeking circuit, but rather part of the processing that establishes the encoding of the inputs to the map-seeking circuits.

The application of some logic to endstop, depth and boundary attributes leads naturally to a discussion of amodal completion of figures with various depth relationships to occluding figures. Bakin, Nakayama and Gilbert [10] demonstrate that response of V1 and particularly V2 cells to pattern inputs beyond their RFs is a more than trivial function of disparity and contrast relationships. While these effects may be the result of horizontal processing, they may instead be another manifestation of forward and backward signals establishing mapped extents modified by very local combinatorial processing of depth, endstop and other attributes.

Virtual Contours

Another behavior that emerges from the circuit characteristics already described is the ability to decompose target images which do not correspond to whole memorized percepts but instead to a set of either hard-wired or learned components. This behavior is demonstrated in Fig. 7-5.

The Kanizsa pattern Fig. 7-5(a) is the raw test image. It is filtered into two components of vertical and horizontal orientation (not shown). Since the

---

[10] Bakin, Nakayama, Gilbert 2000

Kanizsa effect is largely independent of the virtual figure produced, the patterns used to decompose it, Fig. 7-5 (b,c), are in effect co-linearity detectors. [11] The decomposition of the test image by the memorized co-linearity detectors locates the "subjective contours," which are shown as *rbkwd* activity for the two components, Fig. 7-5 (d,e). There are weak co-linearities created by the segments of the disk perimeters which align with the orientation filters. The small responses to these can be seen. The backward mapped alignment of the co-linearity detectors with the original image is seen in Fig. 7-5 (f). Note that this process of decomposition may provide an explanation for location of "subjective contours," but it offers no explanation of the brightness effect for which Kanizsa figures are also famous.

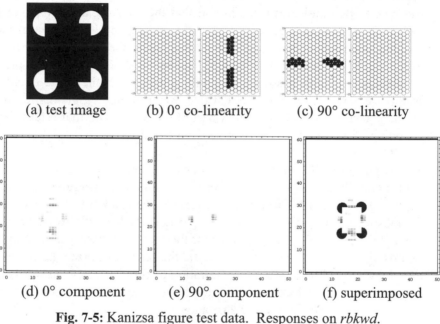

    (a) test image   (b) 0° co-linearity   (c) 90° co-linearity

    (d) 0° component   (e) 90° component   (f) superimposed

Fig. 7-5: Kanizsa figure test data. Responses on *rbkwd*.

---

[11] This demonstration was implemented with the neuronal circuit, so neither scaling nor rotation were available Consequently the co-linearity search patterns were designed to have approximately the correct span for the image being tested In circuits such as those demonstrated in Chapters 2 and 3 a single such co-linearity search pattern could operate over a wide range of scalings and rotations.

Kinetic Depth Effects and Limb Motion Signature Recognition

A number of related psychophysical behaviors beg for unification under the theory proposed here. These are

1. Certain patterns of 2D motion by two related points induce perceived depth effects.
2. Rotational motion of random points on the surface of a transparent cylinder induces a clear impression of the cylindrical shape.
3. A 2D projection of a wire "sculpture" induces a clear sense of the 3D structure if there are sharp angles in the wire.
4. Humans can easily identify the gait and other limb motion signatures of familiar individuals when the motion is defined only by points at the limb joints.

The first three are sometimes termed *kinetic depth effects*. The characteristic common to all these is that rigid line segments are defined, or are assumed to be defined by, pairs of endpoints, and the motion of the endpoints in the viewplane readily induces a sense of the motion of the segment in a 3D space.

First, touching base with evolution, one might ask why the visual system should have such a capability. The obvious answer is that humans, at least, in order to learn the use of their own limbs, need to be able to interpret the movement of others' limbs in 3-space by observing those movements projected in the visual plane. Children have a remarkable ability to mimic complex motion after a single observation, from a single viewpoint. To accomplish this they must map the view plane projection of the motion back into 3-space. Using visual information alone this is an underconstrained problem. But if the structure and motion constraints of a limb are assumed, the projection into 3-space becomes either well-defined, or tractable by selecting familiar or probable interpretations over improbable ones.

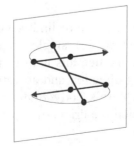

Fig. 7-6: Kinetic depth effect example.

A minimalist case of this occurs in the interpretation of two points moving horizontally in opposite directions, Fig. 7-6. These will often be interpreted as being connected by a rod, so they appear to sweep out arcs, one into and one out of, the plane as they move.[12] The length of the rod is taken to be the initial distance between the points, and in order to maintain this as a constant, the visual system interprets the straight line motion on the view plane to be the projection of two arcs perpendicular to the viewplane which would keep the points related in 3-space by the initially assumed distance.

First one must ask why two independently moving points would be assumed by the visual system to be related at all. It would be less of a stretch to understand the induced perception if it were produced by a line segment stretched between the two points. In the light of the theory here it will be seen that it is the end points of the line segment that are of concern, and that is probably why the primary visual cortex is supplied with end stop detectors, sensitive to ends or sharp inflections in edges.

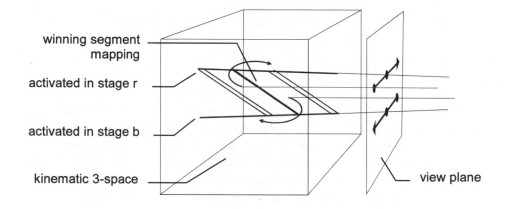

Fig. 7-7: Projection of a line or limb segment from the view plane into the kinematic frame. A single projection into the kinematic frame is shown. In practice this is one of a set of projections determined by viewing direction and an estimation of the orientation of the observed creature's body frame orientation relative to the viewer. The projection is a driven-mapping circuit function.

---

[12] Palmer, p 489-490

If the reader glances back at the forward and inverse kinematics circuit introduced in Chapter 4 it will be obvious why endpoints matter. The kinematics of the limb segment in 3-space (illustrated in Chapter 4 in 2-space) are a mapping from the position of one endpoint to the possible positions of the other endpoint. In that chapter we saw how the candidate positions could be constrained by inhibiting or activating the cells representing those positions in the various layers. The 2D view of the moving endpoints can be directly projected into such a 3D representation, for example Fig. 7-7, by the use of mapping circuits as discussed in Chapter 3. If one considers a snapshot in time, the projection of each of those points is a ray from the observer's eye through the point. In the spatial representation of the kinematics model each point along that ray would be activated and all others inhibited. For a single segment one ray would be activated in memory, or on stage b backward, the other ray would be activated on stage r forward. The mapping chosen must correspond to the assumed length, in this case the observed length at the beginning when the points are at their greatest separation.

When the kinematics circuit, thus initialized, starts to seek mappings between the activated points, it can only choose pairs on the rays, separated by the assumed length (in 3-space) of the segment. There will be two sets of active mappings, one with the top of the segment pointing toward the observer and one set pointing the other way. Within each set, all the segments are parallel at varying distances from the observer's position. Any of these mappings are consistent with the 2D projection of the points, but will be reduced to one either by random variations or some distance and orientation bias skewing the competition between the mappings.

All the circuitry necessary to implement this operation has already been discussed in Chapter 4. The case of limb movement interpretation is simply a multi-segment version of the same process, but a reasonable case can be made that the single segment illusion is a special case use of the multi-segment circuit that must be in place for interpreting viewed limb motion.

It should now be clear why recognition of limb motion signatures from joint position markers is not such a complicated problem after all. If the observer makes the projection of the limb segments into a kinematic frame of reference reasonably well aligned with the owner of the observed limb, the set of discovered mappings for the limb segments directly determine the joint angles, as we have seen in Chapter 4. Interpreted this way, the kinetic mimicry performed by children is based on a rather simple mechanism. Our ability to recognize motion signatures from joint markers is yet another application of the same mechanism. The memory of the gait signature is a recharacterization of the 3-space kinematic interpretation of the joint motion in the viewplane. We apparently store a signature as a sequence of mappings, or more flexibly as a sequence of relationships between mappings. Signature recognition consists of a successful mapping from the currently observed sequence of kinematic mappings to the memorized one – another use of a map-seeking circuit, but one very much related to the tracking behavior demonstrated in Chapter 6.

The other kinetic depth effect illusions can be explained similarly.

Visual Form from Temporal Coherence

In the demonstrations all of the inputs to stage-r from the feature field(s) have been sustained (DC) signals. This allows the inherent oscillatory nature of the circuit to assert itself without being affected by any periodicity in the inputs. It is not necessary (and unlikely in biological systems) for the inputs to have this sustained character. How would the circuit perform if the target in an image were defined by a coherent set of periodic signals while the background was defined by an incoherent set of signals?

This is the condition tested by Lee and Blake[13] in the human visual system. In their experiments, subjects were presented images constructed of random small Gabor patterns of various orientations. A subset of those Gabor patterns, forming a square, was made to flicker or rotate coherently, while the Gabor patterns making up the background were desynchronized. The orientation of the Gabor patterns varied randomly so there were no static edges to be detected by the visual system. Yet the subjects had no problem discerning the square, though defined entirely by the coherence of the wavefront of the signal. This result is considered important because it supports the belief that coherence of cellular activity in the cortices provides the unity of a percept.

In the neuronal circuit, the behavior of individual pathways under oscillatory stimulation of various frequencies relative to their "natural" frequency has been tested so the behavior of the circuit under these conditions can be predicted with high confidence. There are two conditions:

1. The highest frequency among the inputs to the circuit is equal to or lower than the recognition state oscillatory frequency of the circuit;
2. The lowest frequency among the inputs to the circuit is greater than the recognition state oscillatory frequency of the circuit.

---

[13]Lee, Blake 1999

In condition 1 the recognition state periodicity will be governed by the periodicity of the coherent signal evoking the strongest response from a memorized percept. The wave front of the activating signal must be coherent both to evoke a response in stage-m and to activate the r-match cell for that input subfield.

In condition 2 the recognition state periodicity is determined by the coincidence of the rise of the input signal and the fall of the inhibition to stage r forward via the stage r inter-cell. Condition 2 can be considered a sequence of short DC inputs, some of which will be ignored if stage r forward happens to be inhibited when they arrive. Thus as the input frequency rises relative to the natural frequency of the circuit, the latter comes to dominate and the irregularity of the resulting recognition state periodicity diminishes. Kandil and Fahle[14] found that young observers can identify temporally coherent figure and ground stimuli distinguished by delays down to about 10-15 msec. A 10-15 msec delay is long enough to segregate the wave fronts for figure and ground when beat against a 17-25 msec circuit periodicity (assuming the latter isn't a multiple of the former). Under these conditions the wavefronts will not have regular periodicity, but that does not matter. Since a neuronal map-seeking circuit depends on the coherence of the signal fronts within it rather than their regularity, there is no difficulty posed by a recognition state with an irregular rhythm.

Hence, over a wide range of frequencies the coherence of the input signal can define a target and allow the circuit to distinguish it from a non-target, consistent with the experimental results of Lee and Blake, as well as Kandil and Fahle. The interpretation of these results with the map-seeking mechanism finds no contradiction between the two sets of results because it does not rely on any synchrony between the stimulus frequency and some global signal or circuit resonance.

Visual Tracking and Pursuit

If cortical processing can track translation across the retina when no motion information is encoded in the input features (as illustrated in Chapter 6), one might wonder why any cells of V1 would have motion and direction sensitive receptive fields, and, at least in cats, why the majority would have such characteristics.[15] The author puzzled over this until having the opportunity to observe a feral cat hunting in high grass. One aspect is particularly striking: the cat does not fixate the target and track by pursuit but rather stares in

---

[14] Kandil, Fahle 2001
[15] Hubel, Wiesel 1959

the direction of the target with no visible movement of its head or eyes until sufficient displacement of the target causes him to re-center the target in the field of view with head movement, sometimes anticipated by a small saccade.

Assuming that the cat is re-centering using visual and not auditory localization, this means he is tracking the translation of the image of the target across his retinas by cortical processing. The cat's view of the target through the grass is highly occluded. This is an environment in which the clutter is all edges, and the target may present none. In high grass the velocities produced by motion of the prey will be anything but consistent The slope of a snout moving horizontally between two reeds will give a vertical, not horizontal velocity: a naturally occurring variant of the barber pole illusion. The cat's propensity to stare makes velocity sensitive detectors particularly effective in characterizing the image of prey in these circumstances. To its staring eye, movement by the target will define its boundaries with motion vectors, even if those vectors are not consistent. So for a cat in its natural hunting ground motion detectors may be more useful as edge detectors than static edge filters.

In this interpretation the output of velocity sensitive detectors in V1 is a means of distinguishing and characterizing the boundaries of a target rather than providing direct inputs for the calculation of its retinal velocity. Using a map-seeking circuit, speed and direction can be easily and reliably determined by the progression of translation mappings tracking those motion defined boundaries. The barber pole problem shows that this is a far simpler solution than trying to compute global motion from local boundary motion vectors.

7.4 Visual Deficits Associated with Schizophrenia

It must be noted that the author was completely unaware of any visual deficits associated with schizophrenia (or much else associated with it, for that matter) until several years after the behavior of the neuronal circuit was well understood. The awareness arose from a chance reading of a report in *Science News* concerning deficits in backward masking tasks.[16] These were immediately comprehensible to the author in terms of the theory offered here, for similar "deficits" had been encountered in tuning both the idealized neuronal circuit and its electronic realization, and could be recreated at will. The author then searched the experimental abstracts for related manifesta-

---

[16] *Science News*, Vol. 156, Nov 13, 1999

tions, and a list of deficits and physiological manifestations was assembled. All had a single root explanation in terms of the theory.[17]

The following visual deficit categories were drawn from the first search of an abstracts database, and are certainly not comprehensive.

1. Deficits in identifying a target in a backward masking task. Those with the deficit require a longer interval before presentation of the mask to identify a target.[18]
2. Repetition blindness. Those with the deficit show decreased performance in identifying repetitions within rapidly presented visual word lists.[19]
3. Deficient processing of velocity information leading to smooth pursuit abnormalities.[20]
4. Limited visual search and scan path abnormalities.[21]

Seen in the light of the theory all of these abnormalities could result from a failure of relevant map-seeking circuits to converge to a single mapping or at least delay in doing so.

1. In backward masking tasks, if a mask is presented before the circuit is able to converge to a single mapping the recognition state is never achieved, and in most cases a group of memory cells will still be responding. The combination precludes recognition of the pattern presented prior to the mask
2. Recognition of repetitions is an inherent property of the circuit, as demonstrated above. However, the failure to converge to active mappings will result in a failure to locate the repetitions, since the mapping is the locating signal.
3. The translational mappings provide the velocity signal of a target before pursuit stabilizes it on the retina, and they provide the correction signal if the target changes velocity or the pursuit is inaccurate. Again, failure to converge quickly to a correct mapping will deny a clear tracking signal.
4. Identification of possible saccade targets is probably a manifestation of attention shifting to candidate features identified at low resolution on the

---

[17] A number of evoked potential findings which might be considered to be supportive were excluded because they require associating surface recordings with the activity of specific cell groups.

[18] Green, Nuechterlein, Breitmeyer, Mintz 1999

[19] Park, Hooker 1998

[20] Chen, Nakayama, Levy, Matthysse, Palafox, Holzman 1999

[21] Kurachi, Matsui, Kiba, Suzuki, Tsunoda, Yamaguchi 1994

periphery via the magnocellular pathway. Two possible failures of the circuit could interfere with this: a) a failure or delay in establishing a mapping on the feature, b) a failure to suppress already identified features to allow an attention shift. The latter could be due to a disruption of the mapping of stage b backward onto stage r backward, or due to a failure of the suppression circuitry.

A Single Point of Failure: Inhibitory Imbalance in Mapping Competition

All the failures are explainable by a single dynamic: insufficient or excess inhibition in one of the stages of the r-match competition circuitry. Prior to encountering the clinical accounts, these failures had been observed by the author in mis-tuned circuits and always resulted either in slow establishment of the proper mapping or a partial or complete failure to do so. The usual manifestations are a delay in establishing the recognition state oscillation (gamma) or the establishment of an irregular oscillatory pattern with a few gamma frequency cycles interrupted by slower irregular periods. At the extreme the slow, regular non-recognition state periodicity is established. The three recordings of Fig. 7-8 illustrate normal and abnormal inhibition cases. (Recordings from the CMOS implementation of the circuit are used here for the clearer display due to smaller time steps and more flexible graphics software.)

Fig. 7-8 shows delay in establishing the recognition state oscillatory response (circuit gamma) in the abnormal cases, Fig. 7-8 (b,c), relative to the normal case, Fig. 7-8(a). The recognition state frequency for the CMOS implementation is about 0.6 MHz. The normal case converges in 10usec (equivalent to 150 msec at 40 Hz). The first abnormal case fails to establish gamma until about 50 usec (equivalent to 750 msec at 40Hz). The second abnormal case fails to establish gamma within the 60 usec test (equivalent to 900 msec at 40Hz). The only difference in the parameters of the three cases was a progressive decrease in the weight of the inhibitory synapse coupling the output of the r-match competition circuit back to each r-match cell.

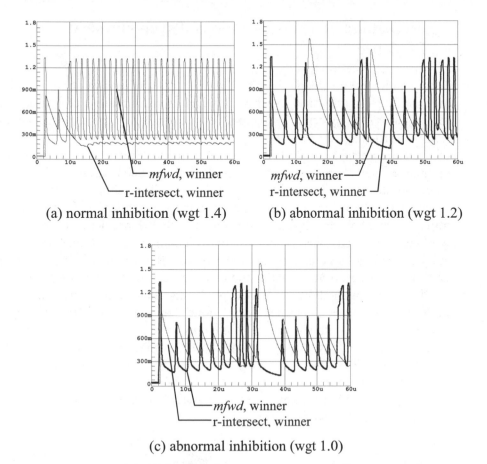

(a) normal inhibition (wgt 1.4) (b) abnormal inhibition (wgt 1.2)

(c) abnormal inhibition (wgt 1.0)

Fig. 7-8: Circuit convergence anomalies.

Insufficiency of inhibition from the competitive circuitry, illustrated above, appears to have a counterpart in schizophrenia. The search which turned up the visual performance deficits also revealed that much suspicion has already fallen on inhibitory irregularities in connection with these deficits and with other manifestations of schizophrenia. This shows up in GABA[22] and PET[23] measurements of activation.

---

[22] Schwartz 1990
[23] Taylor, Tandon, Koeppe 1997

7.5 Repertoire of Transformations: Psychophysical Evidence

To the author's knowledge there has been no comprehensive attempt to determine by psychophysical experiment which transformations are present in the human visual system. Individual researchers have directed their attentions to particular transformations, and so it is necessary to assemble a sense of the experimental results from various sources. Some facts are so commonplace they are frequently used for demonstrations of visual characteristics in children's science museums. The classics of this genre demonstrate the visual system's limited ability to recognize even very familiar inverted faces. NASA, confronted with the problem of astronauts working with each other at all possible orientations, has studied the difficulty of reading facial expression under these circumstances.[24] At the other end of the spectrum very subtle studies have been undertaken to determine the limits of human ability to recognize objects under translation[25] and rotations into the plane. There has been a long debate over whether the mind builds internal 3D models which it attempts to match or whether it recognizes rotations into the plane from one or a combination of previously learned views. This discussion does not take a position in that debate since the mechanism proposed in this book is applicable to both approaches. The question being addressed now is which visual plane transformations need to be implemented for various tasks.

(a)　　　　　　　　(b)　　　　　　　　(c)

Fig. 7-9: Effects of transformations on facial images.

The simplest demonstration illustrates that some transformations preserve identity while others alter it. The familiar face of Audrey Hepburn, Fig. 7-9(a), remains Audrey under perspective transformation, Fig. 7-9(b), but is

---

[24] NASA Ames Research Center, Perceptual and Behavioral Adaptation Group
[25] Dill, Edelman 1997

altered into a far less sublime person, Fig. 7-9(c) by a mild non-linear trans-formation, Fig. 7-10.

Fig. 7-10: Non-linear transformation used on facial image.

While this demonstration is not intended to suggest we recognize faces ro-tated into the plane by using only plane transformations, the fact that a 2D image of a face remains identifiable under a dramatic perspective transfor-mation but not under a mild non-linear planar transformation suggests that we solve the problem of recognizing faces by very subtle measurements.

What emerges is that we are not equipped to recognize as many transforma-tions as we could be. Starting from the premise of this book, that the visual system is built of circuits in which mapping is explicit and fixed (or learned slowly) it should hardly be surprising that evolution has not equipped us as a mathematician designer might have done. Assuming mappings are learned, we learn the ones we need. Assuming mappings are genetically encoded we are descendants of creatures with the most useful sets of mappings for the natural environment. If inverted faces appear before us, we generally know to whom they belong without having to inspect them too closely. But right side up we are pretty fussy about accepting only linear transformations, it would appear. Non-linear transformations seem to imply difference, not equivalence, in faces at least.

Has evolution been chary with transformations because they are very expen-sive in neural terms? If the map-seeking architecture of Chapter 5 is at all indicative, then the costs do not seem prohibitive. The neural investment in a map-seeking circuit is dominated by the r-match and r-intersect cells. If we use the parvocellular pathway as the example, then one can come up with a rough feature field diameter of about 200-300 features in a number of ways (1-2 minute foveal resolution, optic nerve diameter fiber count/centers per feature). From this we have about 50,0000 inputs to *rfwd* of layer 1 and about 28,000 translations, assuming translations for every position that al-lows a full mapping onto *bfwd*. Assuming 5° increments of rotation in the upper two quadrants and 10° increments in the lower, we need 54 rotations. With scaling factors from approximately 1.1 (from 0.95 to 1.05) near unity to 1.2 at the extremes, we need about 40 scalings. For perspectivities, the range needed is sharply constrained by the 5.2° field of view of the fovea. This is equivalent to about a 400mm telephoto lens on a 35mm camera. In this narrow a field of view perspective is single axis and dominated by fore-shortening.[26] Since the perspective can be applied at any angle by rotation, 30 perspective mappings are sufficient for quite fine discrimination of perspective change from motion.

If all of the non-translation mappings are accommodated in layer 2, then 65,000 mappings are needed. For the two layers about 100,000 mappings, requiring 2 major cells per mapping, bring the neural budget to 200,000 cells. Round this to 250,000 for the complete circuit. Finally, assuming each data path is implemented with a redundancy of half a dozen, as suggested in the last chapter, the budget rises to about 1.5 million cells per circuit. For the low resolution retinal periphery, the budget is far lower, and permits generous multiple channels as discussed earlier. In any event, the budget is hardly prohibitive even if there are many hundreds of such circuits in the visual system. And with three layers, the opportunity to compose perspectivities and scalings, for example, permits either a much wider array of transformations, or reduces the neural budget, or both.

So, assuming the neuronal model of Chapter 5 has some validity, we may be more limited by evolution than neural resources in our transformational repertoire. The theory here suggests an interesting set of psychophysical experiments to determine which transformations we are born with, which we can learn, and which we can't. For example, do we really map all perspectivities? Consider Fig. 7-11.

---

[26] Multi-axis linear perspective is therefore a characteristic that would only be present in the internal image assembled from many fixations, and it is questionable whether it is represented even there. Even trained artists rarely draw accurate perspective without construction, and when the subject is composed of curves it is difficult to get it right even with construction.

Fig. 7-11: Test of mapping acuity.

The perspective projection on the right is consistent with only one of the "canonical" views on the left. Which one? Even with the cue of the surrounding box, which is valid in both cases, it is difficult to see which interior figure on the left produced the figure on the right. Yet both appear to be valid projections consistent with the projection of the surrounding box. In the terms of the theory one can conclude that our visual systems appear to have mappings for this perspectivity, but that they are of low enough resolution (or of limited enough extent) that it is impossible to determine which of the two is the consistent projection. In fact, despite appearances, both perspective figures are projections of the lower "canonical" figure.

Fig. 7-12: Comparing projections.

But comparing the true projections, Fig. 7-12, of both the upper and lower "canonical" figures of the previous example, a quite different behavior becomes evident. The lower figure now appears at first glance to be a slightly flatter projection of the upper figure, perhaps just 5°-10° flatter, and slightly tipped to the right. Our visual system makes a strong effort to relate the two figures by some transformation. In this case it is a subtle transformation that involves small adjustments in perspective in two axes, as seen in Fig. 7-13. The transformation of the outer square indicates the mapping of lower into the upper figure of Fig. 7-12. Though the figures are in fact not derived from each other, but simply projections of two different original canonical figures, the visual system ignores this inconsistency in the attempt to relate the majority via a single mapping.

Fig. 7-13: Mapping to match incompatible figures.

The difference in interpreting the two figures above is exactly what one would expect from an evolutionary interpretation of the theory. The mapping between the two patterns of Fig. 7-12 is one that would be encountered frequently in terrain interpretation from viewpoint displacement. Not surprisingly our visual system makes a concerted effort to apply it. In contrast, the need to establish an accurate mapping between a canonical and an extreme perspective view occurs rarely, if ever, in nature. It is the consequence of a man-made environment. Students of architecture (as the author was) are well aware of the liberties that can be taken in making extreme perspective projections of elevations (the architect's term for canonical views) without offending the eye of any but a trained observer.

This is only one example of the kind of psychophysical experiment implicit in the theory advanced in this book.

7.6 Recognition of Assembled Views

When the details that distinguish one object from another, or that distinguish changes in a single object over time, require resolutions that cannot be accommodated in a single view, then the visual system must assemble a composite image of the object from multiple fixations. (A simple demonstration example appears in Appendix I.) Studies of foveal "parsing" of faces show that visual attention is lavished on certain features, such as eyes, while the majority of areas are passed over. The resulting composite image must have a wide range of effective resolutions, and there is some indirect evidence that it is not manipulated as a single image entity. If that is true, it may well be that it cannot be manipulated as a single image entity because it is not stored that way. In the light of this theory some of the most compelling bits of evidence for a linked composite representation come from studies performed at NASA[27] to determine the limits of recognizing airport maps and aerial photos in unfamiliar orientations. Their studies show that presentation of the same map in orientations increasingly remote from the initial presentation either increases the time to recognition, or for some individuals results in a failure of recognition. This strongly suggests that the internal representation of the map has no access to *en masse* transformation, but that the pieces have to be individually transformed and reassembled, which some individuals can accomplish and some cannot.

The difficulty in recognizing the subtleties of facial identity or expression in unfamiliar orientations suggests that the internal data structures used to hold faces may also lack access to *en masse* transformation. The demonstration of Fig. 7-9 indicates unambiguously that a mild non-linear transformation of Audrey Hepburn's face altered its identity. But what transformation was actually responsible for the loss of identity?

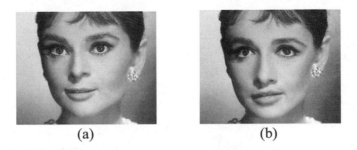

(a)   (b)

Fig. 7-14: Other transformations on facial images.

---

[27] NASA Ames Research Center, Perceptual and Behavioral Adaptation Group

Fig. 7-14, above, provides a clue. In Fig. 7-14(a) Audrey's eyes and mouth were clipped from the original image and pasted into the location of the features in the transformed face. It now takes a much closer inspection to realize that some of the transformations remain. As one would suspect from the foveal parsing patterns, most of the identity of a face resides in its eyes and mouth. The most obvious flaw in Fig. 7-14(a) is the nose, which was left from the transformed image. So a weaker sense of the identity comes from the nose, at least in Audrey's case. Moods are most strongly expressed in the conformation of the eyes and mouth. Recognizing variations in the moods of someone one knows very well, e.g. a parent or child, is far more critical than recognizing someone one knows less well, so the investment in fine discriminatory capability directed at eye and mouth transformations shouldn't be surprising. In Fig. 7-14(b) the transformed features were pasted back into the original face. It is very clear that the identity resides primarily in the eyes and mouth. Audrey Hepburn was chosen as the example precisely because her features are not individually extraordinary, but the regularity and proportion are. Yet, even in her case, eyes and mouth make the identity.

Seen in the light of the theory, one might predict that there is a repertoire of mappings applicable to the visual and conformation transformations of eyes and mouths specifically and that the resolution in these is extremely fine. Meanwhile, the general layout of the features is stored with much lower precision, and subject to much cruder transformations. From these assumptions one might postulate an architecture of two tiers of circuits for face recognition. A very simplified block flow is shown in Fig. 7-15.

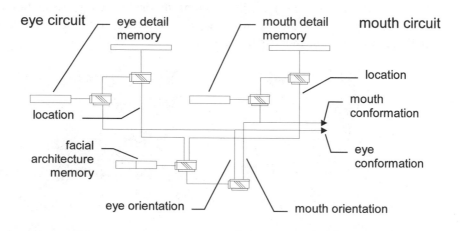

Fig. 7-15: Simple face recognition architecture. (See Fig. 4-9 for symbology.)

The first tier might have separate circuits for eye and mouth features, each storing its own kind of feature in great detail and equipped with the visual and conformation transformations appropriate to that feature. The output of these circuits is the set of mappings which locate the feature and identify the visual transformations and conformation transformations. The location and orientation mappings constitute a recharacterization which is itself captured as facial architecture by the second tier circuit. These can either be stored as a series of orientation views, or used to construct one or more 3D models. The conformation mappings constitute recharacterizations to use as input for a variety of cognitive purposes, such as mood determination and assistance with speech recognition in the case of mouth conformation.

This two-tier architecture explains why we can recognize that we are seeing a face, and its crude characteristics, regardless of orientation, but lose the ability to discriminate mood and fine identity outside a range of rotations. It explains why we interpret the expression of inverted faces composed of upright features as being the mood conveyed by the features in their normal, not inverted, context. This striking demonstration is a staple of visual phenomena exhibits at children's science museums. The architecture also explains why caricaturists can exploit feature similarities yet place them in very distorted facial arrangements without inhibiting recognition.

Chapter 8

Practical Map-Seeking Circuits and Map-Seeking Computers

8.1 Implications of a Map-Seeking Computing Technology

The primary purpose of this book is to introduce a principle and examine its relevance to both biological and machine vision. The use of map-seeking circuits in machine vision can take two directions. It can be a component of a dedicated piece of hardware: an *embedded* application. It can also be the basis of a general-purpose computing architecture far better suited than conventional computers for exploring problems which involve multiple sequences of inverse mapping operations. Map-seeking circuits are to such a computing architecture what the arithmetic-logical unit (ALU) is to conventional computer architecture. **A computing architecture whose primitive operation is the discovery of mappings instead of scalar arithmetic is such a vast leap in scale that it cannot but alter the approach to cognitive processing**. Once that leap is comfortable and natural it will become clear to what degree even our concept of cognitive processing has been strangled by the limitations of conventional computing.

While there has been some discussion of the scale and practicalities of a biological realization of map-seeking circuits, nothing has been said about the feasibility of technological realization. The pace of development of technologies capable of implementing map-seeking circuits threatens to make any sober analysis of upper limits obsolete in short order. Instead the intention of this discussion is to mark a lower bound, based on currently available implementation technologies.

8.2 A Sense of Scale: Digital Implementation

We are now in a position to quantify a question raised in the first chapter. If this is a valid theory of vision, what does it imply about the scale of computation in the visual system? For a first answer to this we will return to the algorithmic circuit which is dominated by dot products, scalar-vector products, and sums. If these are suitably rearranged almost the entire algorithm can be implemented using multiply-accumulate (MAC) operations.

Roughly, the maximum number of MACs required to execute a single iteration in a circuit of three layers plus memory is

$$2 \cdot i_1 \cdot j_1 + 3 \cdot i_2 \cdot j_2 + 3 \cdot i_3 \cdot j_3 + 2 \cdot k \cdot j_3$$

where

i_L is number of mappings in layer L
j_L is dimension of *bfwd* of layer L
k is number of memory locations

The number of operations to compute the non-linearities is small compared
to those needed to compute the dot and scalar-vector products so they are
ignored here.

Using the three-layer circuit from the demonstration in Appendix I as an example of a useful-scale circuit, we will assume

$i_1 = 14.4 \times 10^3$
$j_1, j_2, j_3 = 25600$
$i_2 = 240$
$i_3 = 20$
$k = 100$

and find that the maximum number of operations in each iteration of the algorithm is about 760 million MACs. In digital implementations, because
mappings are eliminated so rapidly during early iterations, about 75% of the
operations can be bypassed during each convergence. For 20 iterations per
frame and 10 frames per second, the algorithm requires hardware that can
execute about 38 billion MACs per second.

To put this in perspective, at least one major vendor of field programmable
gate arrays (FPGAs) produces a chip capable of providing 10-20 billion
MACs/sec when configured to execute a variant of the algorithm described
in Chapter 2. In this device the speed-limiting factor is the routing bandwidth of data on the chip to execute non-translational mappings. The scale-
limiting factor is on-chip memory capacity, which is sufficient to support
fields somewhat smaller than assumed above. The arithmetic and I/O bandwidths of the FPGA are not limiting factors. Using somewhat smaller fields
and pruning active mappings by employing various optimizations, the device is capable of processing 10 to 30 frames per second. A gate array or
application specific (ASIC) implementation of the same architecture could
be built to execute at the required speed with the larger data fields and number of mappings assumed above.

172

Looking forward, routing bandwidth will continue to be the limiting factor. The difficulty, of course, is in the data access pattern for the mappings. Translational mappings are implemented by adjacent shifts in two axes and pose little problem in the current technology. Rotational and perspective mappings however require a large number of concurrent all-to-all routing paths on the chip to keep the arithmetic logic busy. For an 80 element diameter *bfwd/bbkwd*, the routing paths and logic accessible in current FPGA technology provide bisection bandwidths of approximately 5 Gbyte/sec. (As a point of reference, a 5 Gbyte bisection bandwidth is typical of small supercomputers, but the chip has far lower latency.)

Other data access issues involve scaling, scalar distribution and multipoint interpolation if required. Most of these data access issues have solutions very specific to the characteristics of the implementation hardware, so discussion is outside the scope of this book. They are mentioned as a reminder that although the algorithmic circuit is dominated by multiply-accumulate operations, that is not the entirety of the design problem. As this discussion suggests, an efficient digital implementation is necessarily highly specialized.

If a practical two- or three-layer circuit can be implemented in current commercial technology, then it is not far-fetched to assume that within a decade smaller, faster processes and chip stacking technologies will make it feasible to implement a number of map-seeking circuits inside a single package. The implication of this, remarkable as it seems, is that digital technology is approaching the power, if not the compactness, required for what we have estimated to be biological scale cortical computation.

8.3 Map-Seeking Computers

There are numerous applications for which a single map-seeking circuit provides the backbone of a solution. Many tracking applications require fewer mappings and less memory than we have been using in the calculations above. Collision anticipation, as it might be employed in a motor vehicle, requires no target recognition. The objective is to identify a sequence of scale and position mappings for each moving region in the input frames. The map-seeking circuit need only provide the transformation sequences. The rest of the calculation is readily performed on a conventional microprocessor. Military target acquisition and tracking does involve recognition, and therefore requires target sample memories and a larger repertoire of mappings, but is essentially a single-circuit problem. A number of common industrial computer vision applications such as pick-and-place or conveyor monitoring are similarly suited to single-circuit solutions.

Autonomous robots, however, cannot be realized without something approaching a true visual system. The block diagram of a simple terrain interpretation system, Fig. 4-10, contains a good number of circuits operating in map-seeking mode, as well as some in driven-mapping mode. Even if the driven-mapping functions were replaced with conventional processing, the amount of computation required by the seeking-mode circuits is daunting. The earlier discussion of the processing power needed for a single circuit means that systems of such circuits would tax even the largest conventional supercomputers. By contrast, a specialized high level architecture, built around multiple map-seeking chips could provide far greater aggregate computation at a tiny fraction of the cost, even implemented with current technology.

Map-Seeking Units: The Building Block of A Map-Seeking Computer

If the block diagram Fig. 4-10 represents a typical problem to be set up on a map-seeking computer, then such a machine will need to provide many map-seeking-circuits of various dimensions with different mapping sets, and a rich set of pathways to interconnect the circuits. To provide a platform capable of supporting a variety of applications, as the traditional computer does, this multiplicity of circuits and pathways will need to be partly physical and partly virtual. To emulate a variety of map-seeking circuits in a larger architecture each physical circuit will have to be rapidly reconfigurable in dimension, deployed mappings, communication between mappings in different layers, perhaps non-linearities, thresholds and convergence criteria.

Such reconfigurable hardware elements will be termed *map-seeking units* to distinguish them from the generic circuit. The map-seeking unit (MSU) is to a map-seeking computer what an arithmetic-logical unit (ALU) is to a conventional computer. Early computers had one ALU which was used for every calculation including address indexing. One of the first steps to higher performance computing was specializing ALU functions and adding ever more of them to the architecture. Yet in many architectures, to preserve machine language backward compatibility, only one could appear to be directly controlled by the programmer. Ironically, a great deal of extra hardware needed to be added to make use of several ALUs for speed while preserving the illusion of using only one. Conversely, map-seeking computers will have to virtualize one or a few MSUs into appearing as many map-seeking circuits with different dimensions, mappings and other characteristics in various locations in the signal flow.

Verbally this sounds innocent enough. But consider the number of paths in and out of a single map-seeking circuit, Fig. 8-1.

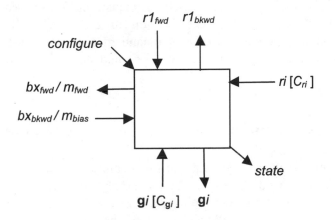

Fig. 8-1: Map-seeking unit interface.

1. $r1_{fwd}$, $r1_{bkwd}$, and bx_{fwd}, bx_{bkwd} (where x is the last layer) are externally accessible forward and backward pathways. If configured with memory mi_{fwd} and mi_{bias} replace bx_{fwd} and bx_{bkwd}. (mi_{bias} is an external gating input to modify mi_{fwd} coefficient.)

2. gi output is the set of mapping coefficients for each layer.

3. ri $[C_{ri}]$ input masks or initializes the r stages of each layer, ri.

4. gi $[C_{gi}]$ input masks or initializes the mappings of each layer, gi.

5. *state* output synchronizes the execution of multiple MSUs.

6. *configure* input sets stage dimensions, mappings and other operating parameters.

Any of the output paths of an MSU may be connected as the source of one of the inward paths of another MSU. To complicate matters the computations of some of these interconnected circuits may have to evolve concurrently while others have to wait until all the circuits providing input have converged. This wouldn't matter if all the circuits in the architecture were physical and analog and interconnected by analog signals. But where map-seeking units have to do multiple duty the connections between the physical units have to be reconfigured on the fly to emulate paths between different virtual map-seeking circuits. This is a well-solved problem: existing crossbar switching technology is more than adequate to implement reconfigurable paths between MSUs. Adequate buffering needs to be available on these paths. The coordination of the data exchange between asynchronously operating MSUs is a standard distributed system design issue with adequate conventional solutions and does not require discussion here.

8.4 Analog Map-Seeking Circuit Implementations

In Chapter 5 it was pointed out that the idealized "neuronal" elements have direct electronic counterparts. In the nearly perfect world of simulated CMOS the translation from the neuronal circuit to transistor circuitry is remarkably direct. This allowed some of the circuit behavior to be more clearly demonstrated using the CMOS Spice simulation because of the finer time step and better graphics.

A simple cell-for-cell, synapse-for-synapse translation of a useful neuronal circuit into analog CMOS is certainly at the frontier of current microchip practice. The speed of CMOS permits a dramatic reduction in number of circuit elements by implementing one dimension of the input field as a shift register. In the simulations the stage-r dimension is $w_{input} \cdot h_{input}$, and stage b dimension is $w_{t \arg et} \cdot h_{t \arg et}$. However, in a CMOS implementation the stage r dimension need only be $w_{input} \cdot h_{t \arg et}$. For each shift a strip the width of the input field and the height of the target memory field is searched, and the entire input field is searched in $h_{input} - h_{t \arg et}$ shifts. To facilitate shifting, stage-r is implemented in $h_{t \arg et}$ parallel rows. Each shift moves the data from one stage r row to the next, and fills the first row from buffer which has been filled from the incoming data stream.

With this reduction an analog chip with an effective input field of 512×512 can be implemented with 12-20 million transistors on a chip of about 1 cm square using a standard 0.35 micron process. The effective speed of the chip is governed by its oscillatory period, and that in turn is limited by capacitance and the practical current limits of the driving transistors. Because of the capacitance and the need for monotonicity, subthreshold operation is ruled out. The trade-offs between speed and current leave the operating frequency of the recognition state at about 500 kHz. Assuming a maximum of 20 cycles for convergence, a shift can be executed each 40 μsec + transfer and settling time: say 50 μsec. 512 shifts (though some of these are priming) require 0.0256 sec. Therefore 39 frames of 512×512 per second can be pushed through the circuit.

The discussion above assumes a single layer circuit in which only translation maps are implemented. Multiple rotational mappings cannot be conventionally implemented because of practical limits of crisscrossing metal layers. Combinations of switching and multiplexing can be used but the solutions are very complex and beyond the scope of this discussion. A general purpose analog MSU must be considered impractical until richer interconnects on a chip become feasible. However, for specific applications with limited mapping requirements an analog or hybrid solution might be justified in an application where the performance requirement justifies the cost of development.

Spice simulation confirmed that the dynamics of an analog CMOS implementation duplicate the dynamics of the neuronal circuit and that performance objectives could be met within space and power budgets. The dominant factor affecting the speed of the circuit are the capacitances of the metal runs across the warp and woof of the stage r and stage b intersections and the gate capacitances of the "synapses" connected to them. For simulation these have been overestimated by a factor of two over the calculated values. The circuit's recognition-state oscillatory frequency is 550-590 kHz, non-recognition state about 220kHz. (See Fig. 5-7.)

The real obstacles to building an analog map-seeking circuit did not show up in the Spice simulation. What the Spice simulation could not mimic is the systematic and random variations in transistor characteristics across the surface of a chip. These variations, in the order of 50% for the submicron processes, would wreak havoc with the required monotonicity, if not compensated for in some manner. This problem also plagues CMOS image sensors, which must introduce corrections for each pixel to keep striations and other variations from appearing in the output image. Adaptive compensation is the only possibility for an analog map-seeking circuit. Just as the CMOS image sensor must calibrate itself from a uniform white field, the map-seeking circuit would have to calibrate its compensation periodically from a uniform wavefront through the circuit.

The second kind of obstacle also could not register in the Spice simulations because they were not large enough. When thousands of signals are contributing to a single superposition, the threshold of the transistor whose gate is receiving the sum is reached when the contributors are each very small: less than half a millivolt. Of course noise becomes an overwhelming problem at this point. The only way around this is to partition the superposition so that it is summed by a tree structure, each node of which has a limited number of inputs. This allows the inputs to reach a reasonable voltage before the adder conducts, minimizing the noise problem. It also addresses another problem. Even though the peak synapse current in the design is about 1 μa, several thousand signals contributing to a single trace produce enough current to induce migration. Partitioning the contributors eliminates this damaging current surge. Unfortunately the cost of this solution is increased chip real estate and transistor count.

A third problem is yield. Transistor killing imperfections are proportional to chip area, so large area dies, such as this, have low yields. Fortunately an analog circuit can tolerate a certain degree of signal drop-out and still function. How much drop-out can be tolerated is not known, so yield cannot be estimated.

The last obstacle is more simply solved. No existing package technology has enough contacts to get the data on and off chip as analog signals. Fortunately A-D and D-A converters of sufficient performance are readily implemented and permit all interchip communication to take place digitally encoded.

Chapter 9

Related Issues and Lines of Future Research

9.1 A Taxonomy of "Recognition" Tasks

It may seem odd that a book on visual cognition should spend so little space on an activity called "recognition." One of the reasons for focussing on the terrain interpretation problem in this book is that it is a well-defined problem: the observer has to build a three dimensional model of the terrain sufficiently accurate to allow him to place his feet without looking.

The term *recognition* is vague until it is applied to a specific task. A human observer may "recognize" a highly made-up actor by nothing more than a mannerism, a gait, or the expression of an eye. If in performing this feat the brain makes use of image data matching, even through a transformation, that is only part of the operation. Far more likely, whatever matching does take place is applied to a derivative, possibly more abstract, representation. But this level of matching is certainly quite different from the matching that takes place when we "recognize" the familial relationship of an ocelot, a lion and a house cat. There are many kinds of recognition, and the fact that we do not have different words to describe them keeps tricking us into hoping for a single operation to span all of them. We "recognize" so easily and automatically that it is only in trying to duplicate the ability that we have come to appreciate how computationally difficult it is. Robustly duplicating any single aspect is so challenging it is not surprising that no taxonomy of the span of recognition operations has evolved. But finally we may have to address the lack.

The data matching operation presented in Chapter 2 is described as *recognition*, but it is a very limited kind of recognition. Objects of life-or-death interest to predator or prey, for example, present different poses, appear in different orientations, and are often partly occluded. One has to assume that evolution discovered the simplest functional solution to the problem first, and that the most evolved systems probably still show evidence of the earliest solutions. In human face recognition the importance attached to particular features is striking, as demonstrated in Chapter 7. It is not unreasonable to assume that the same holds true for many, if not all recognition tasks. So if a matching mechanism is employed in prey recognition, for example, it probably matches an assembled representation composed of features of the prey which are recognizable under a reasonable repertoire of geometric transformations produced by viewpoint and pose variations. For many animal species the projections of distinctive sections of the head and body can be related by simple plane transformations over a useful range of viewing angles (see example in Appendix I). Therefore, if such sections constitute features in an assembled representation, a limited number of features and mappings is needed to recognize the animal in natural viewing conditions. There is experimental evidence that the cortical representation of the profile images of some objects is decomposed into simple, independently encoded features.[1] A decomposition of this sort would be expected if a feature-by-feature recharacterization of the object's image is taking place.

However, computer scientists have been investigating both hierarchical and statistical feature-based representations for decades without producing a robust recognition mechanism. This suggests there is something different going on in the brain. Or it may be that the approach has been correct but has been steered off track by the desire to avoid what appeared to be an intractable computational problem: the determination of correspondence between memorized and observed features under visual transformation. The taxonomy of recognition must span from "x is identical to y" to "x has a feature which reminds me of the corresponding feature in y, and that feature is unique enough that, in the current context, x is probably the same thing as y, or is a close relative." In a natural environment a majority of recognitions are necessarily of the latter type. So feature-based representations are still the most plausible.

There are three elements to this type of recognition. First, usually only a minority of available features is required to establish identity. Second, similarity of a key target feature to a memorized example may be determined in a variety of attributes and at a number of levels of abstraction. Third, in different contexts a given feature may have different utilities in distinguishing the object to which it belongs. The first and second elements argue for a fluid association of the component features rather than a formal hierarchical structure. All three elements suggest that a fixed statistical weighting of the features in an object representation is too brittle. A dynamic, contextually determined decision criterion relating the features to the object is far more plausible.

The Audrey Hepburn images of Chapter 7 provide interesting examples of the elements of recognition just discussed. By most statistical measures the entirety of Fig. 7-14(b) would constitute a match to Fig. 7-9(a), the real Audrey, but human "recognition" reports two different identities. It does so despite the identical facial shape and nose, because the key features used in

---

[1] Tsunoda, Yamane, Nishizaki, Tanifuji 2001

identifying faces, the eyes, cannot be mapped to the eyes of real Audrey by any permissible transformation. Though the eyes of Fig. 7-14(b) were created by a mild transformation of the eyes of Fig. 7-9(a), it is not a transformation our visual system accepts as valid in the context of faces.

To make matters more difficult, we don't always employ the same features to establish recognition of a face, and the limits for acceptable transformation of those features varies widely. Everyone has seen the crudely pixelated image of Abraham Lincoln so frequently used to demonstrate our powers of recognition. The mosaic is recognizable because the shape of the beard and the shape of the head, though highly distorted by the pixelation, evoke the only black-bearded, long-headed man in our common experience likely to be chosen as a visual example: Abraham Lincoln. Had the pixelation been of Ulysses Grant, fewer of us would be impressed by the power of evocation of a crudely pixelated image. This is a simple demonstration of Bayes' Theorem.

Perhaps "recognizing" pixelated Lincoln is nothing more than data matching to a few memory candidates, one of whose responses is very favorably biased because of a high prior probability. However, when Audrey Hepburn is pixelated, Fig. 9-1, in the same way as Abe, to most observers she is difficult to distinguish from a general class of females with dark hair, dark eyes and a generally broad-cheekboned face. This is true despite what ought to be a very high prior probability bias because we have just been discussing her, and perhaps even looking at the image which produced the pixelation.

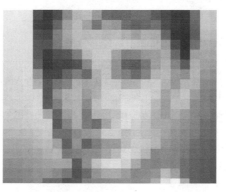

Fig. 9-1: Pixelated facial image.

What makes a feature distinguishing is not at all obvious. We know that if a good artist makes a line drawing of Audrey, we have no problem recognizing her. But if we apply an edge filtering algorithm to Fig.7-9(a) the result, Fig. 9-2, has lost its identity, at least to human observers.

Fig. 9-2: Edge filtered facial image.

We will leave it at this. The first part of the recognition conjecture is that most real world recognitions are keyed on one or a very few features capable of being matched by standard transformation and recharacterization to the corresponding feature or features in the prototype. The second part of the conjecture is that context dynamically restricts the permissible transformations and alters the weighting of key features in the higher level matching processes. Audrey's eyes and mouth in Lincoln's face would probably fail to evoke recognition of Audrey, though the author has not yet tried the experiment. Make-up artists and terrorists undoubtedly know very well the limits of feature recognition in the presence of misleading context.

9.2 All very well, but…

The French have a jesting curse: "That's all very well in practice, but it will never work in theory." The capabilities of map-seeking circuits demonstrated in the previous chapters promise a powerful practical method, but the evidence supporting map-seeking as a cortical theory is incomplete and indirect. It is useful to summarize which aspects of the principle are supported by evidence and which are not.

Some neurophysiological evidence, as discussed in Chapter 7, makes a strong case for an architecture of reciprocal pathways, the backward path carrying a memorized percept. It also argues for an intersection (in the logical sense) of the forward and backward path signals at least in V1 and possibly V2.

Clinical evidence from visual deficits in schizophrenia provides some subtle and specific evidence in support of the map-seeking mechanism. The repetition blindness associated with schizophrenia implies a failure of some mechanism which locates the repeated figures in the input field. A circuit tuned to allow convergence to multiple active translational mappings is a simple, efficient mechanism to implement repeated pattern recognition. Repetition blindness could indicate failure of such a mechanism. Insufficient inhibition, implicated in visual deficits associated with schizophrenia, is readily shown to produce slow or incomplete convergence to correct mappings, resulting in map-seeking circuit performance deficits similar to the clinical visual deficits observed in schizophrenia. Other visual deficits associated with schizophrenia are simply explained by a failure to resolve mappings.

Also as discussed in Chapter 7, much visual psychophysical evidence supports map-discovery as an essential operation of the visual system. This evidence need not be repeated here.

There is no direct evidence of the creation of signal superpositions. Evidence that signal amplitudes diminish as convergence to recognition proceeds, as seen in Fig. 7-1(b), Fig. 7-2(b), is very common in the visual neurophysiology literature. But there are many possible explanations of this and a pruning of the contributors to a superposition is not the most parsimonious of those explanations. As discussed in Chapter 7, the results reported by J.D. Victor indicating a gain in information between 100 msec and 256 msec suggest only indirectly that a pruning of the low information contributors to the signal may be involved.

If evidence of signal superpositions were found, it would imply that some mechanism exploiting the ordering property is used to prune the superposition. Therefore, substantial evidence of signal superpositions would a very important element in solidifying the map-seeking principal as a cortical theory. The nature of the superpositions encountered, if distinguishable, would identify the roles of various cells. Evidence of multiple mappings of an input pattern converging to the recognized pattern would suggest cells corresponding to a *bfwd* path. Evidence of multiple equally primed memory patterns converging to the recognized pattern would suggest *bbkwd* cells, and if transformed geometrically, *rbkwd* cells.

However, detecting a superposition depends on reliable knowledge of the encoding. The working assumption in the neuronal circuit presented here is that signal amplitude is encoded primarily in phase. In this case the pruning of a superposition could only be detected by measuring the shift in phase relative to some pacemaker or clock signal. Major changes in amplitude, as seen in Fig. 7-1(b), would only indicate local gain control. Detection would be simpler for purely amplitude-encoded signals. If the decrease in amplitude seen in Fig 7-1(b), for example, were determined not to be entirely attributable to increasing inhibition, but at least in part due to a decrease in backward signal amplitude, then it could be taken more convincingly as evidence of convergence by pruning superposition components.

In summary, there is strong evidence for a circuit consisting of reciprocal pathways that discovers mappings, and weak evidence that superpositions may play a role. The map-seeking mechanism advanced here does make specific predictions about what signal dynamics ought to be encountered in cortical circuits. But short of elucidating the details of cortical wiring, it will require some subtle experimental strategies to establish conclusively whether a version of map-seeking does or does not occur in the cortices.

9.3 The Main Concepts Revisited

The opening chapter of this book introduced three concepts which have been the backbone of the work presented here. Restating them:

1. The starting assumption is that the vision process is decomposable into a flow of forward and inverse mapping operations. In many of these the application of data patterns at both ends results in the discovery of the correct mapping between them. This discovery mode of the mapping operation, termed *map-seeking*, solves the correspondence problem, which is the key to much visual processing.
2. The discovered mappings themselves, rather than the transformed pattern data, are often used as the inputs for subsequent stages of processing. The use of discovered mappings as data, termed *recharacterization*, is a basis for interpretation and generalization in the visual system.
3. An ordering property of pattern superpositions, apparently unrecognized until now, is the mathematical basis which allows map discovery, or *map-seeking*, to be resolved without combinatorial explosion.

In this book the three have been taken as a unit. In fact they are severable and individually suggest different lines of future research.

Map Discovery, Independent of Means

It has been assumed here that map-seeking is achieved by exploiting the ordering property of superpositions. However, because such a wide range of visual behaviors can be explained by the discovery of mappings whose parameters become inputs for further processing, this would seem to be a vein worth continuing to mine independent of the proposed mechanism.

There are many tantalizing leads. The curvature of a surface can be distinguished with non-planar mappings. Is there a small set of non-planar mappings which allow us to recognize faces in various rotations into the plane? That is to say, given a canonical view and perhaps one or two other views of an individual's face, is there a limited set of brow, nose, mouth, chin and cheek mappings which, assembled, will produce the match between any of a wide range of 2D views and reference views?

Another area for exploration is the applicability of map-seeking to speech and music recognition. The author's first use of neuronal segmentation circuitry was to split pseudo-phoneme sequences into words. The experiment was never extended to the obvious earlier stage of partitioning acoustic input into phonemes. In idealized speech the relationship of the formants maintains a constancy independent of frequency which is a simple translation in the log frequency axis. Variation in speed of speech production ought to be amenable to piecewise or non-linear mapping in the time-axis. The pitch patterns that distinguish question, statement or command in Western languages would be naturally abstracted from content by the progression of frequency mappings along the time axis. Can the sequential dependencies and influences, so common in continuous speech, be accommodated by a standard set of mappings, or by propagating constraints between mappings, as we saw in the inverse kinematics examples? The field of speech recognition is completely outside the author's expertise. Nevertheless, the idea may be worth exploration.

This book has only barely touched on kinematics, kinesthetics and motor control. These appear to be areas in which the essence of the problem is mapping. For example, the problem of adjusting muscular force under dynamic loading to produce the desired trajectory of a limb adds a layer of great complexity to the inverse kinematics problem. Is this also amenable to a map-seeking solution?

Mappings as Data: the Uses of Recharacterization

The use of mappings as data demonstrated in earlier chapters largely falls into two categories. For terrain interpretation the parameters of discovered mappings either directly identify the orientations and displacements of the view patch of terrain, or become inputs to further processing which yields the needed geometric information. In tracking, the mappings characterize both location and orientation of the tracked object. In repeated pattern the mappings also identify locations, and when the repeated pattern itself renders a larger pattern, the location mappings form an abstraction of the larger pattern, independent of the rendering.

In Chapter 4 we saw the discovered mappings representing joint angles in an articulated limb: in other words representing *pose*. And in Chapter 7, a proposed face recognition architecture used one mapping layer to characterize expression or mood transformations in individual features. Earlier in this chapter it was suggested that in a speech recognition application the sequence of mappings that track pitch changes might abstract the mood (in the grammatical sense) of a sentence.

One suspects that these are only scattered examples of useful extractions and abstractions contained in discovered mappings between an input data field and one or more previously captured patterns. To the author's knowledge, the use of mappings as data has never been pursued in a systematic fashion. One of the reasons for attaching the term *recharacterization* to this use is to suggest a class of computational operations. A class invites members, either existing or new ones whose possibility is suspected only because the characteristics of the class call attention to the void. The first instance of recharacterization identified by the author arose from the inherent tracking behavior of a map-seeking circuit with translation mappings. It quickly became apparent that the same circuit could be tuned to identify multiple concurrent repeating patterns in the input field and that the same location-identifying mappings would then recharacterize the arrangement of the repeated patterns. From then on the author made a point of looking at mappings as data for subsequent processing, and all the other uses mentioned above emerged.

One of the author's surmises arising from this work, is that a good part of cognitive processing, both in vision and other modes, is based on sequences of recharacterizations. This is one of the tantalizing leads which the author has not yet pursued.

Superposition Ordering Property

The superposition ordering property is the facilitator. Without it, the whole enterprise is doomed to perish by combinatorial explosion. Exploitation of that property depends, as we have seen, on the monotonicity of the implementation. This poses no problem in an algorithmic mechanism, but cannot be taken for granted in a biological mechanism. Earlier it was mentioned that the process can be made less vulnerable to various imperfections, including violations of monotonicity, by judicious choice of image encodings and more selective combining functions both in r-match and memory dendritic structures. The general principle has been confirmed by experiment, but has not been quantified. A great deal of analysis remains to be done in this area. More work is needed in quantifying the probability of collusion-induced convergence failures. The factors in this analysis should include distribution of image characteristics, types of encodings, selectivity of combining functions and numbers of active memories.

9.4 Conclusion

The lines of investigation suggested above are only a few of those contemplated by the author at one time or another. An idea with powers of explanation and organization, once lodged in the mind, tends to apply itself in unexpected places. Listening to a fugue one suddenly realizes how re-characterization can abstract its structure. Or watching an outfielder start to glide to the eventual landing point of a high fly ball one suddenly realizes he is moving toward the visualized end of a 3-space mapping of a memorized standard trajectory, aligned with and scaled to the velocity vector of the ball off the bat. The map-seeking theory, with each of its elements, seems to alter one's perception of perception in one domain after another. Though this does not prove it is correct, it makes it compelling. Only time will tell if it is correct.

Appendix A

Collusions: Harmful and Otherwise

The following discussion addresses the conditions which can cause a collusion to prevail over a primary target/memory match when the collusion and the valid match are nearly equal. The discussion assumes no competition implemented between memory cells. For convenience we will label the mapped subfields of the input: the target subfield $s_x = d_x(\vec{r})$ matching memory m_x, a collusion-inducing subfield $s_y = d_y(\vec{r})$, and background subfields which we will lump together as S_z.

The subfield s_y is capable of evoking partial responses in n memories, labelled as a group m_Y. We will assume for simplicity that the target subfield s_x evokes no response in any of the m_{Yi} belonging to m_Y. However parts of the subfields s_z belonging to S_z do evoke responses from the m_{Yi} belonging to m_Y, but no single contributor, s_z , to that aggregate S_z has a pattern corresponding to more than a small part of any memory pattern.

The condition is therefore similar to the ones discussed in Chapter 2, except that instead of c3 (which is assumed to be zero) we will look at the effect of input background components c7 and c8.

$$(c1 + c7) \bullet d_x(\vec{r}) \, ? \, (c2 + c8) \bullet d_y(\vec{r}) \qquad\qquad \text{eq. A-1}$$

where

c1, c2, c7 and c8 are partitions of BB as presented in eq. 2-22.

We will assume $n=10$ and that each of the memories m_{Yi} belonging to m_Y have an average of 10 percent of their pattern which matches some part of s_y. The remaining 90 percent of each of their patterns is assumed to be random and not to substantially match any single subfield of the input. We assume $s_x \bullet w_x = 0.9$, and $s_y \bullet w_{Yi} = 0.1$. As in Chapter 2 we will assume $f(z) = z$ for simplicity. (Recall that in practice the non-linearity $f( )$ makes the small contributors to the collusion even less effective.) Therefore the response of each of the $\bar{m}_{Yi}$ must be augmented by other inputs, S_z. That is to say, for each $m_{Yi} \in m_Y$

$$(s_y + S_z) \bullet w_{Yi} > 0.9$$

in order that 10 components m_{Yi} contributing to BB will create a pattern m_Y such that

$$q_y' = s_y \bullet m_Y = s_y \bullet \sum_i \left(f\left(\left(s_y + S_z \right) \bullet \mathbf{w}_{Yi} \right) \cdot \mathbf{w}_{Yi} \right) > 0.9$$

Since $s_y \bullet \mathbf{w}_{Yi}$=0.1 it is obvious that S_z has to contribute significantly in order for the collusion to stay in contention. But if S_z consists of random background elements it is likely to evoke a response from m_x as well as from m_Y. The probability of contributing constructively to m_Y in aggregate is ten times the probability of contributing constructively in the same proportion to m_x. The initial $S_z \bullet \mathbf{w}_{Yi}$ contribution has to be at least 0.9 for the collusion to produce a q_y higher than q_x, the primary match.

As the convergence proceeds the background decays at a faster rate than anything else. Assuming on the first iteration the collusion match q_y is 0.01 ahead of the primary match q_x, the *comp( )* and *f( )* non-linearities will amplify that difference between g_y and g_x to, say, 0.05, assuming the contribution of S_z remains the same. On the next iteration, if the contribution of $S_z \bullet \mathbf{w}_{Yi}$ drops to less than 0.85 the collusion match falls to second place. Consider that the average q_z is very small compared to q_x or q_y. If the initial average q_z is 0.5 (which is unrealistically high), on average the associated s_z are not even a factor on the next iteration, so the associated g_z may decline by nearly the value of k in *comp( )* on the next iteration. Therefore a few very optimal members of S_z are more effective than many average ones in keeping the collusion in the competition.

How likely is such a distribution of S_z? Again, it depends on the image characteristics and encodings (and the combining functions in the dendritic structures). If the binomial random walk discussed in Chapter 2 were the generator of the fields s_z belonging to S_z, the probability is low. For 10 hits in a walk of 100 the probability is less than 10^{-5}. In the real world of highly correlated images the probability is certainly higher. But it should still take many thousands of s_z belonging to S_z to make collusion occur with noticeable frequency.

From the preceding discussion it is apparent there are two conditions propitious for sustaining a collusion. One, which was precluded by assumption for the discussion above, occurs when the field of the primary match s_x itself contributes significantly to the response of m_y.

$$q_y' = s_y \bullet m_Y = s_y \bullet \sum_i \left(f\left(\left(s_y + s_x + S_z \right) \bullet \mathbf{w}_{Yi} \right) \cdot \mathbf{w}_{Yi} \right)$$

The other condition which encourages the success of collusions is a low initial q_x. For example, if $s_x \bullet \mathbf{w}_x$=0.7 initially, on the second iteration g_x will be about 0.3. Now, even without any contribution from s_x, as just discussed, S_z needs to contribute 0.2 to the response of m_Y to keep g_y ahead of g_x. In practice both of these conditions have been present when collusions have been encountered.

Collusion in Test Cases

A variant of the recognition demonstration shown in Fig. 2-3 provides an interesting test of collusion by using a data set in which the number of components likely to collude can be roughly calculated. This test also uses tiles of the valid and distractor memories to create a large number of potentially collusion supporting synthetic distractors. However the tiles are 10 by 10 instead of 20 by 20, so a good portion of 16 are needed to match either the target pattern or a collusion pattern. Five original memories are used to supply tiles, so approximately a fifth of the tiles come from the valid memory. But about half of these contain significant data, so about 10 percent of the tiles available are direct matches of some part of the input image. In a typical test 5000 memories are loaded, of which all but five are constructed of tiles. About 80 percent of the tiles contain at least one tile from the image. The 5000 memories are subjected to 11 rotations (a range of 100° in 10° increments), so *BB* contains 55,000 components initially. However, of the 55,000 components only 80 percent contain tiles of the input, and of these only one rotation makes a good match to the target pattern in the input image. This leaves about 4000 components which are optimal contributors to either the target pattern or a collusion, or perhaps as many as 6000 when augmented by matching tiles from other memories and coincidentally matching rotations of tiles.

Even under these test conditions sustained collusions are not produced unless the initial match between the primary memory and the target is less than optimal, as occurs when target or memory pattern is rotated near 45 degrees from the original orientation (because no mapping interpolation is used in these tests). When a sustainable collusion condition is produced, it can be made predictably unsustainable by reducing the mismatch (e.g. by keeping to orientations nearer the original axes) or reducing the number of tiled distractors.

From this discussion it is also evident why surface interpretation tests, such as those appearing in Chapters 2 and 3, never show any evidence of collusion. Because there is usually one, or at most a few active memories, even large numbers of mappings in two layers cannot produce enough long lived components in *BB* to make a sustainable collusion possible.

Appendix B

Encoding and Pattern Sparsity

This section provides an illustration of the means to achieve different levels of sparsity and their consequences. It treats images whose meaning is carried entirely in their edges: line drawings with uniform black lines on a white background. The principle also applies to other image attributes.

In the first example a very large but finite repertoire of n_1 line segments of various lengths, orientations and curvatures is used to construct an image. To encode this image we will imagine a mechanism which is capable of deciding which of the n_1 line types in the repertoire each segment in the drawing matches, and locating an index or locus point for that segment. The machine has n_1 codes to represent the line types and A loci on which to locate any segment. We will assume for simplicity that no locus can be used more than once. The drawing could be represented in an array of A elements, each of which is an n_1 bit vector (disposing of the savings of encoding the line type in binary). The encoding machine simply identifies each line segment by type and drops the code vector for it in the appropriate element of the array. A drawing constructed from p_1 line segments will be represented by p_1 non-zero vectors of the array, each of which has only one bit turned on. There are $n_1 A$ bits of which p_1 are 1, so the sparsity will be defined as $1 - \dfrac{p_1}{n_1 A}$. We will call this Type 1 encoding. It is the sparsest of the encodings we will consider.

A superposition of k Type 1 encoded images is likely to have about $1 - \dfrac{kp_1}{n_1 A}$ sparsity if there are not a large number of images k relative to A, and so long as the average p_1 is low compared to A. Therefore line elements of the same type from two images are not likely to occupy the same locus in the superposition, so there is no need to count line types in each locus of the superposition. This makes Type 1 a desirable encoding for superpositions. It is also a desirable encoding for memory function. If there are k memories, each encoded with the pattern of one image, and one of the k images is presented to that set of memories, only one will have a full response, and all the rest will have zero or virtually zero response. These virtues make a circuit employing Type 1 encoding almost proof against collusions.

Type 2 encoding will be more frugal. By restricting all the lines segments in the repertoire to a constant short length, but with a variety of curvatures and orientations the original drawing may be closely duplicated. The number of line types in the repertoire, n_2, will be much smaller than n_1 so the total number of memory "cells" n_2A is much reduced from Type 1 encoding, but the number of line segments p_2 necessary to encode the drawing rises by a factor approximately equal to the ratio of the lengths in Type 1 to Type 2. The sparsity $1 - \dfrac{p_1}{n_2 A}$ compared to $1 - \dfrac{p_1}{n_1 A}$ is decreased by the change in both n and p. A superposition of Type 2 images obviously runs a greater risk of having multiple lines segments occurring in the same locus, so now the superposition encoding must accommodate a count in each location instead of a bit. (This will be the case for all subsequent encoding types.) For memory function Type 2 encoding has a related disadvantage because some of the k-1 image memories which are not the match of a target image may nevertheless have some response to it. A circuit using Type 2 encoding begins to have some small chance of collusions.

In Type 3 encoding we will simply restrict the number of line types to just two kinds of short, straight segments at perhaps three dozen different orientations. One kind of line segment encodes lines which are continuations of other lines at both ends. The other kind of line segment encodes lines which cease or continue with an abrupt bend at one end (termed end-stopped). Having lost the large repertoire of curved segments, the drawing must now be a bit cruder, but it can be encoded in much less space. But again the sparsity decreases and the chance of spurious response of memories increase. Collusions become more probable.

In Type 4 encoding we keep the same number of line types but reduce their length. Now the drawing can be finer but the coding becomes even less sparse and the collusions increase further.

In Type 5 encoding we shorten the line to a point. Now there is no orientation information to maintain, and no end-stop information. So each locus in the image needs only one bit of information: the presence or absence of black ink in the drawing. Superpositions must still count multiple bits per locus. Type 5 encoding is obviously the representation used in most of the demonstrations in this book. It encodes with the lowest sparsity, $1 - \dfrac{p_1}{A}$, and is the worst case for collusion.

Type 3 is a subset of the encoding believed to be produced by the mammalian primary visual cortex. The sparsity difference between Type 3 and Type 5 should be kept in mind when interpreting the biological implications of the demonstrations in this book.

Appendix C

Mapping Circuits for Geometric/Trigonometric Calculations

A number of trigonometric calculations are required to convert mapping parameters from the viewplane into geometric characteristics of the terrain patch as projected into the 3D model. These equations would seem to imply that the brain contains neurons that multiply and compute arctangents, sines and the like. In practice these functions are simple enough to be computed by a conventional multilayer "neural net." But even simpler, and more consistent with the rest of what is being proposed, is a subset of the mapping circuits we have already seen. As an example, consider the equation for computing ψ from the derivation in Appendix D1.

$$\psi_{cir} = \tan^{-1}\left(\Delta\psi / \left(yscale - 1\right)\right)$$

A simple map-seeking circuit consisting only of the paths r_{fwd}, b_{bkwd}, and the matching operation can be set up to compute any function of two variables. In this case let the range of $\Delta\psi$ values be represented by r_{fwd} inputs, and the range of *yscale* values be represented by b_{bkwd} inputs. Each g_i represents a particular value of ψ. Each q_i responds just to pairwise mappings between all the $<\Delta\psi, yscale>$ pairs which produce the particular value of ψ represented by the associated g_i.

The algorithmic circuit has a very compact oscillatory neuronal counterpart. It requires one neuron for each *rfwd* path, one for each *bbkwd* path, three for each *rm-ri*, and a few extra to implement the competition. This circuit can compute a continuum of values for ψ by using combinations of adjacent *rfwd* and *bbkwd* activations to encode values intermediate to the values represented by each path. The interconnection pattern can be trained by activating the $<\Delta\psi, yscale>$ pairs and the correct *rm-ri* and using Hebbian learning to set the synapses.

Appendix D1

Inclination, ψ, from Single Axis Foreshortening

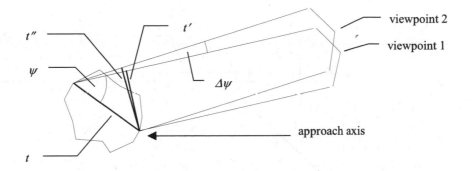

Fig. D1-1

$$\sin \psi = \frac{t'}{t} \ , \ \ \sin\left(\psi + \Delta \psi\right) = \frac{t''}{t}$$

$$yscale = \frac{t''}{t'}$$

$$\frac{\sin\left(\psi + \Delta \psi\right)}{\sin \psi} = \frac{t''}{t'} = yscale$$

$$\frac{\sin \psi \cos \Delta \psi + \cos \psi \sin \Delta \psi}{\sin \psi} = \cos \Delta \psi + \frac{1}{\tan \psi} \cdot \sin \Delta \psi = yscale$$

for small $\Delta \psi$

$$\psi = \tan^{-1}\left(\frac{\Delta \psi}{yscale - 1}\right)$$

Appendix D2

Transverse Inclination, θ, from Two Axis Foreshortening

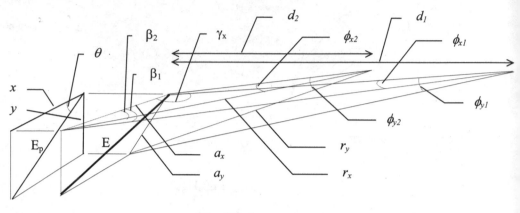

Fig. D2-1

E connects two points on the patch lying generally transverse to the line of approach. E can also mark the extremes of a narrow mapped extent whose long dimension is generally transverse to the line of approach. x, y, E_p are projections of a_x, a_y and E respectively in the line of view, θ is inclination of mapped extent transverse to line of view. Assumptions: d_1, d_2 >>x, y and γ approximately equal for both viewpoints.

$$\frac{a}{\sin\phi_1} = \frac{d_1}{\sin\beta_1} \quad , \quad \frac{a}{\sin\phi_2} = \frac{d_2}{\sin\beta_2}$$

$$\frac{d_1 - d_2}{a} = \frac{\sin\beta_1}{\sin\phi_1} - \frac{\sin\beta_2}{\sin\phi_2}$$

$$\frac{d_1 - d_2}{a} = \frac{\sin(\phi_1 + \gamma)}{\sin\phi_1} - \frac{\sin(\phi_2 + \gamma)}{\sin\phi_2}$$

$$\frac{d_1 - d_2}{a} = \frac{\sin(\phi_2 - \phi_1)}{\sin\phi_1 \sin\phi_2} \cdot \sin\gamma$$

$$\frac{(d_1 - d_2) \cdot (\sin \phi_1 \sin \phi_2)}{\sin(\phi_2 - \phi_1)} = a \cdot \sin \gamma$$

$$\frac{a}{\sin \phi_2} = \frac{r}{\sin \gamma}$$

$$\Delta d = d_1 - d_2 \quad, \quad \Delta \phi = \phi_2 - \phi_1$$

$$r \cdot \sin \phi_2 = \frac{\Delta d \cdot \sin \phi_1 \sin \phi_2}{\sin \Delta \phi}$$

Let x be the projection of a, as shown above. Assuming $r >> a$

$$r = \frac{x}{\sin \phi_2}$$

$$x = \frac{\Delta d \cdot \sin \phi_1 \sin \phi_2}{\sin \Delta \phi}$$

For both x and y axes,

$$x = \frac{\Delta d \cdot \sin \phi_{x1} \sin \phi_{x2}}{\sin \Delta \phi_x} \quad, \quad y = \frac{\Delta d \cdot \sin \phi_{y1} \sin \phi_{y2}}{\sin \Delta \phi_y}$$

$$\tan \theta = \frac{y}{x} = \frac{\sin \Delta \phi_x \cdot \sin \phi_{y1} \sin \phi_{y2}}{\sin \Delta \phi_y \cdot \sin \phi_{x1} \sin \phi_{x2}}$$

$$\theta = \tan^{-1} \left(\frac{\sin \Delta \phi_x \cdot \sin \phi_{y1} \sin \phi_{y2}}{\sin \Delta \phi_y \cdot \sin \phi_{x1} \sin \phi_{x2}} \right)$$

Appendix D3
Relationship of Binocular Disparity and Surface Orientation

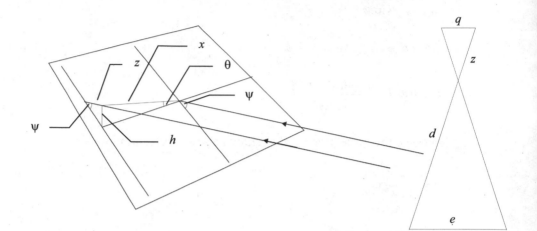

Fig. D3-1

$$\tan\psi = \frac{h}{z}$$

$$\tan\theta = \frac{h}{x}$$

$$z\tan\psi = h = x\tan\theta$$

$$z = \frac{x\tan\theta}{\tan\psi}$$

$$\frac{q}{z} = \frac{e}{d}$$

$$q = \frac{e}{d}\cdot\frac{x\tan\theta}{\tan\psi}$$

Appendix E

Neuronal Circuit Simulation Equations

Each cell computes the current value of its output v_t from a dendritic combining function *dcf* over n dendritic tree regions:

(i) $$v_t = v_{t-1}\left(1 - k_{decay}\right) + k_r \cdot \max\left(\frac{k_1}{1 + e^{k2(D-k_3)}} + k_4, 0.0\right)$$

where

k_{1-4} are shape parameters

k_{decay} is decay time parameter

k_r is gain parameter

$D()$ is defined in (ii)

(ii)

$$D = T\left(dcf\left(regionfn(\mathbf{w}_1 \bullet \mathbf{v}_1', b), regionfn(\mathbf{w}_2 \bullet \mathbf{v}_2', b), ..., regionfn(\mathbf{w}_n \bullet \mathbf{v}_n', b)\right), k_{dcf\_threshold}\right)$$

where

$\mathbf{w_i}$ is the weight vector for region i at time t

$\mathbf{v_i'}$ is the input vector for region i at time t

$k_{dcf\_threshold}$ is the threshold applied to the entire *dcf*

b is a bias input signal

regionfn( )   is defined in (iii)

dcf( )   is defined in (iv)

$T()$ is defined in (v)

(iii)

$$regionfn(x,b) = s_{pol} \cdot \begin{cases} \min\left(0, k_{regthresh}\right) & if \quad x \cdot (1 + s_k \cdot b) < k_{regthresh} \\ x \cdot (1 + s_k \cdot b) & if \quad k_{regthresh} \le x \cdot (1 + s_k \cdot b) < k_{regclip} \\ k_{regclip} & if \quad k_{regclip} \le x \cdot (1 + s_k \cdot b) \end{cases}$$

where

s_k is bias values 0,1 depending on region type

s_{pol} is 1 if excitatory region, -1 if inhibitory region

$k_{regthresh}, k_{regclip}$ are threshold and clipping constants, $k_{regthresh}$ may be negative

(iv)

$$dcf\left(r_1,...,r_n\right) = nodefn(< node\_ex >, < node\_ex >) | < region\_range\_ex >$$

where

$$node\_ex := nodefn(< node\_ex >, < node\_ex >) | r_i | < region\_range\_ex >$$
$$region\_range\_ex := rangesumfn(r_p,....,r_q)$$

> $nodefn(\ )$ is defined in (vii)
> $rangesumfn(\ )$ is defined in (vi)

Thus the *dcf* can be represented as a functional expression which captures a tree organization of node functions.

A simple example:

$$dcf\left(r_1,r_2,r_3,r_4,r_5\right) = nodefn\left(nodefn\left(nodefn\left(r_1,r_2\right),nodefn(nodefn\left(r_3,r_4\right)),r_5\right)\right)$$

(v) $$T\left(x,k\right) = \begin{cases} 0 & if & x < k \\ x & if & x \geq k \end{cases}$$

A subset of adjacent regions is referred to as a *region-range*. The dendritic combining function *dcf* applies one of three classes of function to the output of regions or region-ranges to arrive at a value for the entire ensemble of regions.

1) Regions within a region-range are combined by thresholded summation referred to below as the *rangesumfn*:

(vi) $$rangesumfn\left(r_n,...,r_m\right) = \begin{cases} 0 & if & vsum_{range} < k_{range\_threshold} \\ vsum_{range} & if & vsum_{range} \geq k_{range\_threshold} \end{cases}$$

where

$$vsum_{range} = \frac{1}{n-m+1} \cdot \sum_{i=m..n} r_i$$

$n...m$ are the index bounds of the regions in the range.

r_i is the output of the *i-th* region

$k_{range\_threshold}$ the threshold for the region-range

2) Regions and region-ranges are organized as the leaves of any binary tree. The *dcf* applies one of four dyadic *node functions* at each bifurcation of the tree, yielding a single value at the root of the tree.

$$
\text{(vii)} \qquad nodefn(in1, in2) = \max\left(0, \begin{cases} T\left(in1 + in2, k_{sum\_thresh}\right) \\ T\left(\dfrac{in1 + in2}{2}, k_{avg\_thresh}\right) \\ T\left(in1 \cdot in2, k_{prod\_thresh}\right) \\ adj(in1, in2) \end{cases} \right)
$$

where

$$k_{sum\_thresh} \geq 0$$

$$k_{avg\_thresh} \geq 0$$

$$k_{prod\_thresh} \geq 0$$

$T()$ the threshold function is defined in (v)

$adj()$ is defined in (viii) or (ix)

$$
\text{(viii)} \qquad adj(x, y) = \begin{cases} 0 & if \quad x + y - 1 \leq 0 \\ x + y - 1 & if \quad 0 < x + y \leq 2 \\ \dfrac{x + y}{2} & if \quad 2 < x + y \leq k_{clip} \\ k_{clip} & if \quad k_{clip} < x + y \end{cases}
$$

where the unbiased range of x and y is 0.0 to 1.0.

The *multiplicative regime* is the area $0 \leq x \leq 1, 0 \leq y \leq 1$. Any area outside of this is the *additive regime*. The *adj()* function saturates at a value k_{clip}. The behavior of the circuits is tolerant of the implementation of the *adj()* function, so any continuous function with this character should suffice. A sigmoid implementation of a similar function

$$
\text{(ix)} \qquad adj(x, y) = \max\left(\frac{k_1}{1 + e^{k2(x+y-k3)}} + k_4, 0.0 \right)
$$

where

$$k_1 = 4.0, \quad k_2 = -1.1, \quad k_3 = 2.0, \quad k_4 = -1.0$$

whose surface is shown in Fig. E-1.

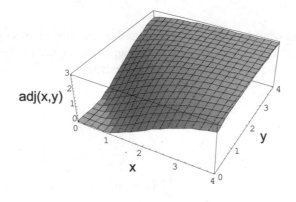

Fig. E-1

Circuits

The connectivity of a circuit is specified by a connection mapping function *cmf* between cell inputs v_{in} and cell outputs v_{out}. v_{in} is identified by <cell, region, synapse> tuples, v_{out} is identified by cell.

(x) $cmf\left[<i,r,s>_{in}\right] \rightarrow i_{out}$

where *i* specifies cell, *r* specifies region, *s* specifies synapse. On each iteration the synapse inputs for each cell are set to the previous iteration value of the cell output to which they are connected.

(xi) $\forall i, \forall r, \forall s: \quad v_{in}\left[<i,r,s>\right]' = v_{out}\left[cmf(<i,r,s>)\right]$

Eqs (i)-(xi) define the recurrence describing the full behavior of a circuit. The identity of a particular circuit lies primarily in its *dcf*s and *cmf*, although there is some variation in decay constants and sigmoid parameters in certain classes of cells.

Learning in Cells

In the circuits discussed only some excitatory regions are capable of learning. Two kinds of learning behavior are possible: episodic and incremental. Only episodic learning is used in the tests reported. Under episodic learning, prior weights of affected synapses have no influence on the new weights. When a cell enters the condition in which learning takes place (generally when D in eq.(i) exceeds a high threshold) one of the following behaviors occurs:

(a) All regions capable of learning reset their weight vector $\mathbf{w}_i$.
(b) All regions capable of learning which have input above some threshold reset their weight vector $\mathbf{w}_i$.

In either case the new weight values are adjusted to reflect the distribution of the input vector $\mathbf{v}_i'$ and normalize the dot product $\mathbf{w}_i \bullet \mathbf{v}_i'$. Under episodic learning

$$(xii) \quad if \quad \sum \mathbf{w}_1 > c_{threshold} \quad then \quad \mathbf{w}_i = \frac{\mathbf{v}_i'}{\sum_{j=1}^{n} (\mathbf{v}_j')^2} \quad i = 1 \ldots n$$

Appendix F
Attention Shift Circuitry

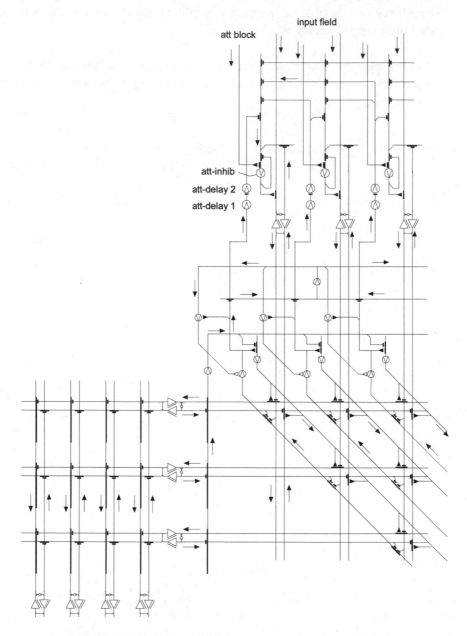

Fig. F-1: Single layer circuit with attention-shifting.
See Fig. 5-5 and Fig 5-6 for signals and symbol definitions.

Fig. F-1 presents the extension to the basic neuronal circuit responsible for shifting attention after a target pattern has been recognized. It is implemented by the *att-delay 1*, *att-delay 2*, and *att-inhib* cells. These use the signals propagated back along *rbkwd* in combination with the output of the successful r-match to suppress input to those *rfwd* cells whose activity has been matched by the responding memory.

The duration for attention on a recognized target is tunable by the activation, thresholds and time constants of the *att-delay 1* and *att-delay 2* cells. The inputs to *att-inhib* are the associated *rbkwd* and the outputs of *att-delay 2* cells in the immediate neighborhood. The dendritic structure of the *att-inhib* requires both the associated *rbkwd* signal AND the activity of several *att-delay 2* in the neighborhood to inhibit the input to the associated *rfwd*. This performs the intersection of *rfwd* and *rbkwd* signals but only blocks those which occur in the neighorhood of other *rbkwd* activity.

The *att block* input provides an external control signal to block attention shifts. The *att-delay 1* and *att-delay 2* cells are tuned so that a number of cycles of pulses at the recognition-state frequency are required for the *att-delay 2* cell to fire. The non-recognition state frequency is too low to push the output of *att-delay 1* high enough to fire *att-delay 2*.

Once a part of the input to *rfwd* is suppressed the remaining inputs compete for recognition. As each pattern is recognized it is in turn suppressed and the recognition proceeds.

Appendix G
Regularity Extraction Circuitry

The regularity extraction operation requires the intersection of *rfwd* and *rbkwd* to be gated onto *rfwd*. This is accomplished by inhibiting all *rfwd* cells except those whose backward pair is active. There are several simple circuit additions which can accomplish this. One is a very simple variant of the attention focus circuitry which already takes advantage of the same intersection property. Another circuit variant simply applies a general inhibition to all *rfwd* cells at a point proximal to the driving activation, but this general inhibition is itself inhibited for each forward cell by the output of its backward pair. The first of these alternatives is shown in Fig. G-1

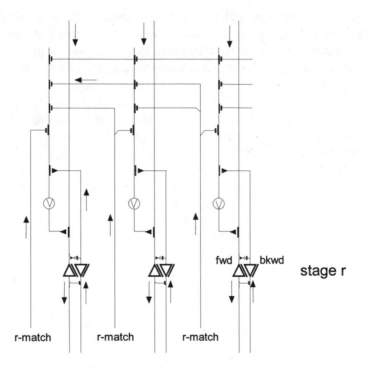

Fig. G-1: Regularity extraction circuitry on stage r.

Appendix H

Three-Layer Circuit Convergence Data

Fig. H-1 shows r_f*fwd* intersect r_i*bkwd* for three-layer circuit used in aerial photo recognition test described in Chapter 2. Test data is shown in Fig. 2-11. The mappings in the three-layer circuit are equivalent to the mappings in the two-layer circuit: 13 rotational mappings in layer 2 and 121 scaling mappings in layer 3 produce composed mappings equivalent to the 1573 layer 2 mappings in the two-layer circuit. Compare Fig. H-1 to Fig. 2-12.

In each row the intersection data for layers 1, 2 and 3 for the same iteration are arranged vertically. The results of the mappings at each layer are evident: translation in layer 1, rotation in layer 2 and scaling in layer 3.

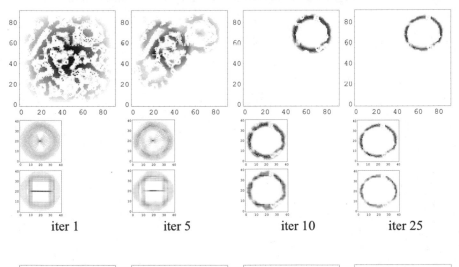

| iter 1 | iter 5 | iter 10 | iter 25 |

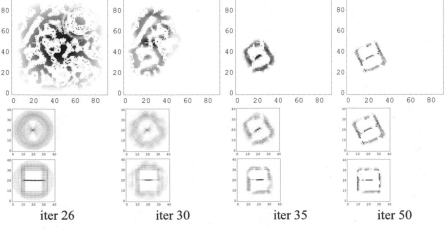

| iter 26 | iter 30 | iter 35 | iter 50 |

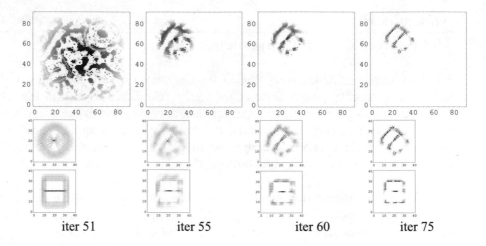

iter 51   iter 55   iter 60   iter 75

Fig. H-1: Three-layer circuit $r_i fwd \cap r_i bkwd$ data

Appendix I

Recognition by Parts

A simple demonstration of "recognition by parts," or recognition of an assembled view, uses a distinctive search pattern (in this case a pair of generic cat eyes) to locate a potential target pattern[1], and then with a sequence of attention shifts searches for other expected elements of the target (generic cat snout, left and right ears). The latter elements are sought in a limited range of translational positions predicted using the mapping parameters of the locating pattern and a model of the configuration of the target animal. The spatial characteristics and limited dimensions of the elements allow a limited repertoire of mappings to match a roughly 50-degree range of head poses despite differential displacements due depth differences.

The circuit used in this demonstration has three layers (translation, rotation, x and y scaling), each with four feature fields corresponding to oriented edge operators centered on 0, 45, 90 and 135 degrees. The input field is 200 × 200 × 4, the search pattern is 80 × 80 × 4. For clarity and compactness the edge data shown below is formed by superimposing the four feature fields. The table after the figures shows the location, orientation and scaling parameters for each of the search patterns. (Note that the search patterns are not clipped from the target image.) The initial locating pattern may be one of a number of concurrently active memories (tested with up to six distractors.) Subsequent expected-element memories are activated singly.

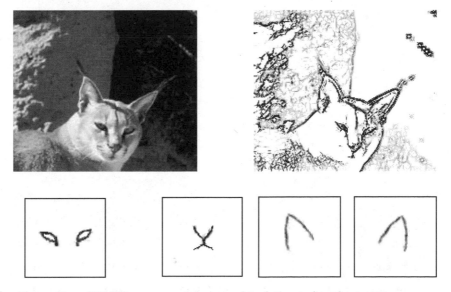

locating pattern (80x80)   patterns sought relative to locating pattern

---

[1] Bobcat photo courtesy U.S. Fish and Wildlife Service

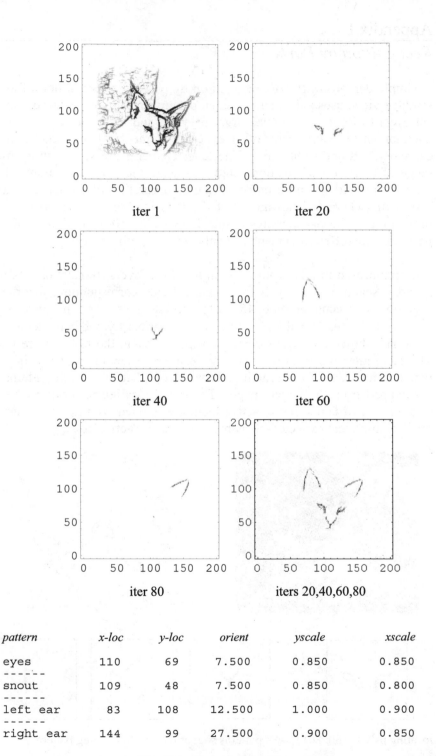

pattern	x-loc	y-loc	orient	yscale	xscale
eyes	110	69	7.500	0.850	0.850
------					
snout	109	48	7.500	0.850	0.800
------					
left ear	83	108	12.500	1.000	0.900
------					
right ear	144	99	27.500	0.900	0.850

Fig. I-1: Recognition by parts.

Bibliography

Agmon-Snir, H., Segev, I. (1996). The Concept of Decision Points, in Bower, J.M. (Ed.) *Computational Neuroscience, Trends in Research 1995* (41-46), San Diego: Academic Press.

Arathorn, D.W. (2001). Map-Seeking: Recognition Under Transformation Using A Superposition Ordering Property. *Electronics Letters* 37(3):164-165.

Bakin, J.S., Nakayama, K., Gilbert, C.D. (2000). Visual Responses in Monkey Areas V1 and V2 to Three Dimensional Surface Configurations. *Journal of Neuroscience* 20(21):8188-98.

Blackwell, K.T., Vogl, T.P., Alkon, D.L. (1998). Pattern matching in a model of dendritic spines. *Network: Computation in Neural Systems* 9: 107-121.

Chen,Y., Nakayama, K., Levy, D.L., Matthysse, S., Palafox, G., Holzman, P.S.(1999). Psychophysical isolation of a motion-processing deficit in schizophrenics and their relatives and its association with impaired smooth pursuit. *Proceedings of the National Academy of Sciences* 96(8):4724-29.

Dill, M., Edelman, S.(1997). Translation invariance in object recognition. A.I. Memo No 1610, MIT.

Douglas, R.J., Martin, K.A.C.(1990). Neocortex, in Shepherd, G.M. (Ed.) *Synaptic Organization of the Brain* (389-438), New York: Oxford University Press.

Gray, C.M., McCormick, D.A. (1996). Chattering Cells: Superficial Pyramidal Neurons Contributing to the Generation of Synchronous Oscillations in Visual Cortex. *Science* 274:109-113.

Graziano, M.S.A., Cooke, D.F., Taylor, C.S.R. (2000). Coding the Location of the Arm by Sight. *Science* 290:1782-86.

Green, M.F., Nuechterlein, K.H., Breitmeyer, B., Mintz, J. (1999). Backward masking in unmedicated schizophrenic patients in psychotic remission: possible reflection of aberrant cortical oscillation. *American Journal of Psychiatry* 156(9):1367-73.

Heath, M., Sarkar, S., Sanocki, T. and Bowyer, K.W. (1997). A Robust Visual Method for Assessing the Relative Performance of Edge Detection Algorithms. *IEEE Transactions on Pattern Analysis and Machine Intelligence* 19 (12), 1338-59.

Hubel, D.H., Wiesel, T.N. (1959). Receptive fields of single neurons in the cat's striate cortex. *Journal of Physiology* (London), 160:106-154.

Janssen, P., Vogels, R. and Orban, G.A. (2000). Selectivity for 3D Shape That Reveals Distinct Areas Within Macaque Inferior Temporal Cortex. *Science* 288:2054-56.

Julesz, B. (1960). Binocular Depth Perception of computer generated patterns. *Bell Systems Tech Journal* 39:1125-62.

Kakei, S., Hoffman, D.S., Strick, P.L. (2001). Direction of action is represented in the ventral premotor cortex. *Nature Neuroscience* 4(10):1020-25.

Kandil, F.I., Fahle M. (2001). Purely temporal figure-ground segregation, *European Journal of Neuroscience*, 13:2004-2008

Kaufman, S.A. (1993). *The Origins Of Order*, New York: Oxford University Press.

Keverne, E.B. (1999). GABAergic neurons and the neurobiology of schizophrenia and other psychoses. *Brain Research Bulletin* 48(5):467-73.

Koch, C., Segev, I. (2000). The role of single neurons in information processing. *Nature Neuroscience* 3(12):1171-76.

Kohonen, T. (1988). *Self-Organization and Associative Memory, Second Edition*, Berlin: Springer-Verlag.

Kurachi, M., Matsui, M., Kiba, K., Suzuki, M., Tsunoda, M., Yamaguchi, N. (1994). Limited visual search on the WAIS Picture Completion test in patients with schizophrenia. *Schizophrenia Research* 12(1):75-80.

Lades, M., Vorbruggen, J.C., Buhmann, J., Lange, J., von der Malsburg, C., Wurtz, R.P. and Konen, W. (1993). Distortion Invariant Object Recognition in the Dynamic Link Architecture. *IEEE Transactions on Computers* 42:300-311.

Lee, T.S., Nguyen, M. (2001). Dynamics of subjective contour formation in the early visual cortex. *Proceedings of the National Academy of Sciences* 98(4):1907-1911.

Lee, S.-H. Blake, R. (1999). Visual form created solely from temporal structure. *Science* 284:1165-68.

Marr, D. (1982). *Vision*, New York: W.H. Freeman.

Marr, D., Poggio, T., (1977). Cooperative computation of stereo disparity. *Science* 194:283-287.

Marr, D., Poggio, T., (1979). A Computational Theory of Human Stereo Vision. *Proc of Royal Society of London* B 204:301-328.

Merryman, R.F.K., Cacciopo, A.J.(1997). The optokinetic cervical reflex in pilots of high-performance aircraft. *Aviation Space and Evironmental Medicine* 68:479-87.

Morrone, M.C., Tosetti, M., Montanaro, D., Fiorentini, A., Cioni, G., Burr, D.C. (2000). A cortical area that responds specifically to optic flow, revealed by fMRI. *Nature Neuroscience* 3(12):1322-28.

NASA Ames Research Center, Perceptual and Behavioral Adaptation Group, Online abstracts.

Palmer, S.E. (1999). *Vision Science*, Cambridge: MIT Press.

Park, S., Hooker, C. (1998). Increased repetition blindness in schizophrenia patients and first degree relatives of schizophrenia patients. *Schizophrenia Research* 32(1):59-62.

Pascual-Leone, A., Walsh, V. (2001). Fast Backprojections from the Motion to the Primary Visual Area Necessary for Visual Awareness. *Science* 292:510-512.

Prazdny, K. (1980). Egomotion and Depth from Optical Flow. *Biological Cybernetics* 36:87-102.

Saffran J., Aslin R.N., Newport E.L. (1996). *Statistical Learning by Eight-Month-Old Infants*, Science 274:1926-28.

Science News, *Schizophrenia May Involve Bad Timing*, Nov 13, 1999, vol.156:309.

Schwartz, B.D.(1990). *Early information processing in schizophrenia.* Psychiatric Medicine 8(1):73-94.

Stevens, K.E. (1981). *The Information Content of Texture Gradients.* Biological Cybernetics 42:95-105.

Supèr, H., Spekreijse, H., Lamme, V.A.F. (2001). *Two distinct modes of sensory processing observed in monkey visual cortex (V1).* Nature Neuroscience 4(3):304-10.

Taylor S.F., Tandon, R., Koeppe, R.A. (1997). *PET study of greater visual activation in schizophrenia.* American Journal of Psychiatry 54(9):1296-98.

Tsunoda, K., Yamane, Y., Nishizaki, M., Tanifuji, M. (2001). *Complex objects are represented in macaque inferotemporal cortex by the combination of feature columns.* Nature Neuroscience 4(8):832-38.

Ullman, S. (1979). *The Interpretation of Structure from Motion.* Proceedings of the Royal Society of London B 203:405-26.

Van Essen, D.C., Anderson, C.H, Olshausen, B.A. (1994). *Dynamic Routing Strategies in Sensory, Motor and Cognitive Processing,* in Koch, C., Davis, J.L. (Eds.) Large Scale Neuronal Theories of the Brain (271-99), Cambridge: MIT Press.

Victor, J.D. (2000). *How the brain uses time to represent and process visual information,* Brain Research 886:33–46.

Zhou, H., Friedman, H., von der Heydt, R. (2000). *Coding of Border Ownership in Monkey Visual Cortex.* Journal of Neuroscience 20(17):6594-6611.

Index

2.5D model (Marr), 72